Economics of Unconventional Shale Gas Development

NATURAL RESOURCE MANAGEMENT AND POLICY

EDITORIAL STATEMENT

There is a growing awareness to the role that natural resources, such as water, land, forests and environmental amenities, play in our lives. There are many competing uses for natural resources, and society is challenged to manage them for improving social well-being. Furthermore, there may be dire consequences to natural resources mismanagement. Renewable resources, such as water, land and the environment are linked, and decisions made with regard to one may affect the others. Policy and management of natural resources now require interdisciplinary approaches including natural and social sciences to correctly address our society preferences.

This series provides a collection of works containing most recent findings on economics, management and policy of renewable biological resources, such as water, land, crop protection, sustainable agriculture, technology, and environmental health. It incorporates modern thinking and techniques of economics and management. Books in this series will incor porate knowledge and models of natural phenomena with economics and managerial decision frameworks to assess alternative options for managing natural resources and environment.

The Series Editors

More information about this series at
http://www.springer.com/series/6360

William E. Hefley • Yongsheng Wang
Editors

Economics of Unconventional Shale Gas Development

Case Studies and Impacts

Springer

Editors
William E. Hefley
Katz Graduate School of Business
University of Pittsburgh
Pittsburgh, PA, USA

Yongsheng Wang
Washington and Jefferson College
Washington, PA, USA

ISBN 978-3-319-11498-9 ISBN 978-3-319-11499-6 (eBook)
DOI 10.1007/978-3-319-11499-6
Springer Cham Heidelberg New York Dordrecht London

Library of Congress Control Number: 2014955366

Printed on acid-free paper

Springer is part of Springer Science+Business Media (www.springer.com)

Contents

Introduction

William E. Hefley, Megan K. Kiniry, and Yongsheng Wang

Abstract Shale gas development has changed the energy discussion in the United States, as existing reserves of natural gas coupled with horizontal drilling and hydraulic fracturing make exploitation of these reserves economically feasible. US energy portfolio has changed significantly due to this new development. The importance of natural gas is seen as likely to continue to expand over the coming years and is expected to increase even further with environmental considerations, such as greenhouse gas emissions. This chapter provides a background view of unconventional shale gas development across the United States.

Emerging Importance of Unconventional Shale Gas Development

Technological advances in horizontal drilling and hydraulic fracturing (or "fracking") make it economically feasible to exploit the vast reserve of shale gas and oil in the United States. It significantly changed the existing energy portfolio. With the supply of cheap gas, related industries have started building plants near those new low-cost energy sources. New supply chains and industries started forming in these areas. This new development increased competitiveness of US manufacturing and stimulated its export. The importance of natural gas is seen as likely to continue to expand over the coming years and is expected to increase even further with environmental considerations, such as greenhouse gas emissions (MIT Energy Initiative 2011).

W.E. Hefley (✉)
Joseph M. Katz Graduate School of Business and College of Business Administration,
University of Pittsburgh, Mervis Hall, Pittsburgh, PA 15260, USA
e-mail: wehefley@katz.pitt.edu

M.K. Kiniry
College of Business Administration, University of Pittsburgh,
Pittsburgh, PA 15260, USA
e-mail: mkk30@pitt.edu

Y. Wang
Department of Economics and Business, Washington and Jefferson College,
60 South Lincoln Street, Washington, PA 15301, USA
e-mail: ywang@washjeff.edu

W.E. Hefley, Y. Wang (eds.), *Economics of Unconventional Shale Gas Development*, Natural Resource Management and Policy 45,
DOI 10.1007/978-3-319-11499-6_1

Horizontal drilling and hydraulic fracturing producing natural gas from deposits such as the Marcellus Shale is making the United States a net producer of natural gas, rather than being a net importer of natural gas (Natural Gas Weekly, July 19. 2010). In fact, studies have estimated the recoverable reserves in just the Marcellus Shale at over 489 trillion cubic feet (Tcf), making the Marcellus Shale the world's second-largest reserve, with only the South Pars field in Qatar and Iran being larger (Engelder 2009). With the Marcellus Shale deposits sitting deep below 95,000 square miles in New York, Pennsylvania, West Virginia, Ohio, Maryland, and Virginia, this huge gas deposit is physically close to the population centers of the Mid-Atlantic and Northeast United States. An existing and potential market of over 16 billion cubic feet (Bcf) of natural gas per day resides within a 200-mile radius of the Marcellus Shale deposits.

The spread of Marcellus drilling in Pennsylvania has increased rapidly in recent years. Figure 1 shows the number of unconventional wells permitted and drilled in Pennsylvania between 2004 and 2013. Projections suggest that as many as 60,000 Marcellus wells will exist in Pennsylvania by 2030 (Hopey 2011).

The predicted natural gas output from shale is predicted to be higher than estimated earlier because of a significantly larger number of drilling rigs producing new wells and faster production times (i.e., more wells drilled per drilling rig resulting in faster cycle times to gas sales) (Pursell 2010). In Pennsylvania by the middle of 2011, there are more than 1,600 Marcellus Shale wells in production, producing 432 billion cubic feet (Bcf) of natural gas during the first half of 2011 (Olson 2011). Marcellus Shale well production in southwestern Pennsylvania alone, including Allegheny, Armstrong, Beaver, Butler, Fayette, Greene, Washington, and Westmoreland counties, during the first 6 months of 2011 increased 55 % to 127 billion cubic feet (Litvak 2011).

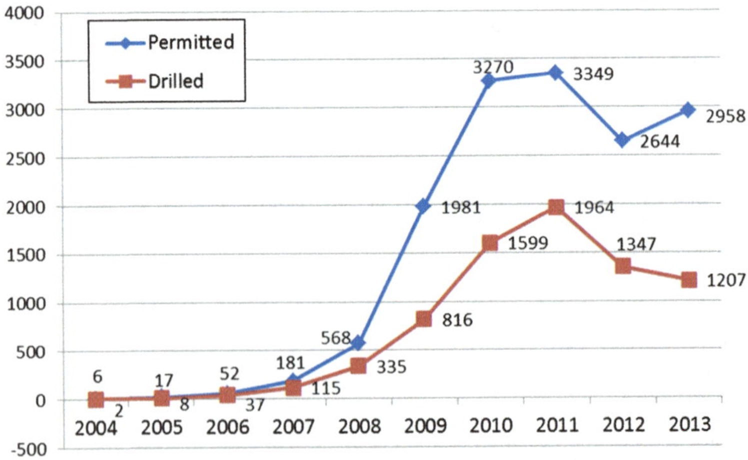

Fig. 1 Number of unconventional wells permitted and drilled in Pennsylvania (Source: PA Department of Environmental Protection, Bureau of Oil and Gas Management, Oil and Gas Reports) (http://www.portal.state.pa.us/portal/server.pt/community/oil_and_gas_reports/20297)

These technological innovations in America's natural gas industry are resulting in new opportunities for growth in domestic production, employment, government revenue, and savings of millions in energy costs from other countries. Specifically, production of "shale gas" has become highly widespread and growing throughout the United States and developing globally. With many countries looking to reduce their dependence on imported natural gas through the development of natural gas from shale, this growth is not limited to the United States. In 2014, the UK government is planning to award licenses for onshore shale gas exploration (Williams 2014). The application of these technologies for shale gas exploration and production causes rapid transformation to the environmental and socioeconomic landscape also which continues to develop every day; therefore, stakeholders must now address the impacts and challenges that result from the application of these innovations in unconventional shale gas development.

Shale has become a highly publicized word throughout the news. Publicized, yet unproductive as 50 % of surveyed Americans noted little to no knowledge of fracking with 60 % with no opinion on the subject (Boudet 2014). An uneducated public is unable to participate in the constructs of a national growth opportunity; therefore, the following research has been accumulated to bridge the gap between recognition and understanding.

Previous studies have examined the economic impact of exploration and production in the mining sector. For example, Black et al. (2005) found that an earlier coal boom spurred economic growth in the non-mining sectors, while the subsequent coal bust resulted in lower economic growth in the non-mining sectors of the region.

As Fig. 1 shows, Marcellus drilling in Pennsylvania has risen significantly since the first well just over a decade ago. As the number of Marcellus wells continues to grow and the awareness of this industry becomes better understood (both for positive and potentially negative impacts), there have been a number of studies that are examining the economic impact of the Marcellus Shale development. Several of these studies address the economic impact of Marcellus Shale drilling (e.g., Considine 2010; Considine et al. 2010; Barth 2010; Higginbotham et al. 2010; The Perryman Group 2008), while others examine the environmental and social impacts (Sample and Price 2011; Ubinger et al. 2010; U. S. Department of Energy 2009).

Beyond the direct spending impacts of Marcellus plays, discussed further in Chap. 2 by Hefley and Seydor, there are additional economic impacts that come as a result of this spending. Kay argues that these impacts may be mixed; some will be winners, while others may not (Kay 2011). These impacts extend throughout the entire supply and value chains of the Marcellus Shale wells, as explained by Kathryn Klaber, former president and executive director of the Marcellus Shale Coalition, who described economic impact as not just coming from drilling and exploration. In an interview with the Pittsburgh Post-Gazette (Gannon 2010), she said, "It doesn't stop with the natural gas companies. There are law firms, accounting firms, small town grocers and dry cleaners all starting to realize – in the areas where this is happening – that there is business to be had and economic opportunities throughout the supply chain."

These additional impacts are comprised of indirect impacts and induced impacts. The indirect impacts are additional economic activity of the value chain network caused by the economic activity of the direct industry. The induced impacts are additional economic activity of all other unrelated firms and households caused by the economic activity of the direct impacts and the indirect impacts. Examples of these ripple effects in Marcellus Shale economic activity are further described by Considine (2010), in his economic impact analysis for the American Petroleum Institute:

> For example, drilling companies hire trucking firms to haul pipe, water, and other materials to a well site. This trucking firm in turn must buy fuel and other supplies to supply these services and hire drivers to operate the trucks. The truck suppliers in turn acquire goods and services from other firms, such as repair shops, parts distributors, and other suppliers. So Marcellus investment sets off a business-to-business chain of spending throughout the economy. These economic impacts are known as indirect impacts. When the drivers go out and spend their paychecks, that spending stimulus sets in motion a similar chain reaction, known as induced impacts. For example, the driver spends his new income on fishing and hunting that stimulates local bait and tackle shops, convenience stores, and other establishments.

The developments of unconventional shale gas development present both opportunities and challenges. A study on shale gas development of Southwestern Pennsylvania by Environmental Law Institute and Washington and Jefferson College in 2014 states that

> The industry has created jobs, generated wealth for some property owners, and after the passage of a state impact fee, provided local governments with a new source of revenue. At the same time, the rapid development of these resources has raised questions of management and planning for local governments in the Commonwealth, including consideration of socio-economic, health, environmental, and economic impacts (Environmental Law Institute and Washington and Jefferson College 2014).

Education will encourage stakeholders to make more informed decisions about the progression of drilling presently and towards a long-term future. This remains an issue that has many facets. Farmers are concerned about the viability of their organic crops, residents are concerned about their homes and cities, and others worry about water and other possible environmental concerns, such as pollution and contamination from drilling, fracking, and gathering (Kretschmann 2014; Crompton 2013; Mufson 2014; Beaver 2014). If successfully managed, there could be reduced dependence on foreign resources, lessened pollution, and a strengthened economy. Balanced policies will provide a sustainable development for the shale gas regions. The key is to convince all parties to sit down and have an open mind to discuss and find solutions. It is apparent that the current technology is not possible for us to fuel all cars without direct or indirect support from traditional energy sources of fossil fuel and not possible to power all homes with 100 % renewable energy. In fact, according to the Energy Information Agency (EIA), renewable energy accounted for less than 10 % of all energy consumption in 2013, and more than 80 % energy was still provided by fossil fuels.[1] It would take a con-

[1] Energy Information Agency, http://www.eia.gov/beta/MER/index.cfm?tbl=T01.03#/?f=A (Retrieved on 7/30/2014).

siderable amount of time to change energy portfolio of a country even without technological hurdles. Not to say, we all understand that wind is not blowing all the time; not all states are as sunny as Florida and Arizona; land has a limit to produce industrial level of crops for biofuel; and nuclear is famous for its super long building cycles. So, how can we transition from fossil fuel to renewable energy? With the significant amount of reserve, shale gas can serve as the source of transitional fuel. Natural gas provides less carbon emission than other fossil fuels such as coal and oil. It can keep wind farms and solar power farms running when there is no wind and cloudy. It can replace petroleum gas and fill up vehicles. It also can replace coal in the power generation. These changes of energy sources will transform the structure of the economy. Policy and resources have to be ready to accommodate related labor force transition and training.

Hydraulic fracturing and horizontal drilling technology make it commercially viable to recover natural gas and oil from shale. Production has created an affordable addition to the Nation's energy supply, while being a proven effective stimulation technique. Without these advanced technologies, it has been estimated that the United States would lose 45 % of domestic natural gas production and 17 % of our oil production within 5 years (American Petroleum Institute 2014). Through the spread of hydraulic drilling, the United States has risen as the chief producer of natural gas with a forecast to succeed as the paramount oil producer by 2020 (Smith 2013). This achievement has supported the creation of over 600,000 jobs, which is estimated by the National Association of Manufacturers to increase towards 1,000,000 by 2025. In 2012, unconventional oil and gas development, mostly from shale, supported 1.7 million jobs, with unconventional gas development accounting for over 900,000 of those jobs in 2012, and projected to grow to over 2 million jobs in unconventional gas development by 2035 (IHS 2012). The American Chemistry Council determined that a 25 % increase in the supply of ethane (a liquid derived from shale gas) could add even more jobs; provide billions in federal, state, and local tax revenue; and spur billions in capital investment. IHS Global Insight estimates that development of shale gas resources will add $926 of disposable household income annually between 2012 and 2015, and the amount could increase to $2,000 by 2035.

These impactful results are due to the large-scale use of the process known as "fracking," typically used in conjunction with horizontal well shafts. The procedure includes the blending of water, sand, and chemicals known as pumping fluids. These fluids are injected under high pressure and then directed down and across unto drilled wells into the shale layer, as far as 10,000 feet below the surface. The pressurizing of materials causes the underground rock layer to crack. These cracks remain open due to mixture's sand, which allows the natural gas to flow up from the shale. The additive chemicals are incorporated to assist with the gas and oil flow. Chemicals account for less than one percent of the mixture, with fracturing water and sand accounting for more than 99 %. Roughly 200 tanker trucks are needed to deliver more than a million gallons of water for fracturing. In 2010, the US Environmental Protection Agency estimated that 70–140 billion gallons of water are used to fracture 35,000 wells in the United States each year (Earthworks 2011). This amount of water is exponentially great with the power to support the annual water consumption of 40–80 cities each with a population of 50,000 (FracFocus 2010). Concern has been attributed to the environmental impact of hydraulic fracturing; therefore, a flowback recycling process has been introduced. An example of flowback recycling is found in Marcellus where fluid used in fracturing is brought aboveground to be collected for reuse in a new well. The flowback and produced water is typically stored on site in tanks or pits before treatment, disposal, or recycling. Recycling of flowback water reduces demand for freshwater and lessens wastewater disposal.

Within the past decade, the combination of hydraulic fracturing with horizontal drilling has opened up shale deposits across the country and brought large-scale natural gas drilling to new regions (Earthworks 2011). This process is also used to continue use of aging wells, and experts believe 60–80 % of all wells drilled in the United States in the next 10 years will require hydraulic fracturing to remain operating (FracFocus 2010).

A prevalent example of the use of these processes is the spread of Marcellus drilling in Pennsylvania, which has increased rapidly in recent years. Between 2005 and 2007, 161 wells were drilled in Pennsylvania. In 2008, this number more than doubled with 332 Marcellus wells drilled in Pennsylvania. Drilling almost doubled again in 2009 with 816 wells drilled in Pennsylvania. The number doubled again in 2010 with 1,599 Marcellus wells drilled in the Commonwealth (DEP 2014). By earlier this year, Pennsylvania Department of Environmental Protection records show that 2,773 wells have been drilled into the Marcellus Shale and almost 7,500 permits have been issued, with projections suggesting that as many as 60,000 Marcellus wells will exist in Pennsylvania by 2030 (Hopey 2011). The predicted natural gas output from shale is predicted to be higher than estimated earlier because of a significantly larger number of drilling rigs producing new wells and faster production times (i.e., more wells drilled per drilling rig resulting in faster cycle times to gas sales) (Pursell 2010). In Pennsylvania, by the middle of 2014, there are more than 7,600 Marcellus Shale wells in production (DEP 2014). Marcellus Shale well production in southwestern Pennsylvania alone includes Allegheny, Armstrong, Beaver, Butler, Fayette, Greene, Washington, and Westmoreland counties.

Other US Department of Energy identified shale reserves including the Antrim (Michigan); Barnett (Texas); Caney (Oklahoma); Conasauga (Alabama); Eagle Ford (Texas); Fayetteville (Arkansas); Floyd (Alabama); Gothic (Colorado); Haynesville (Louisiana); Collingwood-Utica (Michigan); New Albany (Illinois, Indiana, Kentucky); Pearsall (Texas); Chattanooga, Ohio, and Marcellus Shales (Pennsylvania, New York, Ohio, West Virginia); Utica (Pennsylvania, New York and Ohio); and Woodford (Oklahoma) shales. Other countries having significant shale gas reserves elsewhere in the world include China, Argentina, Mexico, South Africa, Australia, Canada, Libya, Algeria, Brazil, Poland, and the United Kingdom.

Dramatic results have been possible through robust exploration, production, and hydraulic fracturing. Many other regulations address land use, safety, traffic, and other potential impacts of shale energy development. States have a long and successful history of regulating oil and gas activities. State regulators continually review their regulations through collaborative efforts with industry and also with public-private partnerships like FracFocus, the State Review of Oil and Natural Gas Environmental Regulations (STRONGER), and the Groundwater Protection Council (GWPC).

Horizontal drilling and hydraulic fracturing producing natural gas from deposits such as the Marcellus Shale is making the United States a net producer of natural gas, rather than being a net importer of natural gas. In fact, studies have estimated the recoverable reserves in just the Marcellus Shale at over 489 trillion cubic feet (Tcf), making the Marcellus Shale the world's second-largest reserve, with only the South Pars field in Qatar and Iran being larger.

To uncover the extended miles of shale, developed infrastructure is required. The development in the Northeast is estimated to push towards $80 billion in infrastructure growth through 2035 with $70 billion attributed to the Marcellus play (Novak 2014). This development predicts a thriving natural gas market for the future. A thriving future with progression and growth will require the subsistence of jobs, process enhancement, and serviceable economic capital.

Evaluating shale development in respect to national values is linkage for both experts and the inexperienced. The blending of news cannot be the sole funnel of education during a time of advancement. With critical decisions pending in legislation and development, evidence must be weighed on the economic potential of unconventional shale development.

Hydraulic fracturing regulation is managed at a federal level, but states have dominance in regulatory efforts. States' regulatory differs throughout the country. Chapter 8 by Murtazashvili addresses the differences in regulatory climate between New York State and the Commonwealth of Pennsylvania. Even down to the local level, there are regulatory issues with approval of shale-related operations (KDKA 2014; Mufson 2014). The extension of fracking into unexplored territories with local residents unfamiliar with resource extraction regulation, such as near the Marcellus Shale, is predicted to be heightened. In a recent ruling, state courts in New York have ruled that local regulations, specifically township zoning regulations, can be applied to limit shale gas development (Taylor and Kaplan 2014). States continue to work towards balancing the economic benefits of accessing new resources with protecting public health, drinking water, and the environment.

Hydraulic fracturing technology has a strong environmental track record and is employed under close supervision by state, local, and federal regulators (Energy From Shale 2013). Compared with the burning of coal, hydraulic drilling machinery reduces greenhouse gas emissions by half with lessened concerns of emissions of mercury or heavy metal, with no decrease in power generation. Additional supervision has also been given to the small percentage of chemicals used in the recent overturn of Act 13 in Pennsylvania. The new legislation requires drillers to provide the state with a list of chemicals used during hydraulic fracturing. While all development has challenges, hydraulic fracturing does not introduce new or unique environmental risks to exploration and production operations, but concerns have been raised due to the potential scale of operations where this technology is applied. Many of these concerns are genuine, and the oil and natural gas industry recognizes that there needs to be a bigger conversation about the development process and the steps being taken to ensure safe operations. Of Pennsylvania's collected impact fees, over $225 million dollars collected in 2013, significant payments are going back to local communities to support local communities, including activities such as environmental enhancement and conservation programs (Public Utility Commission 2014).

While contributing to these programs, the procedure of hydraulic drilling shelters much of the environment's resources through its underground operations including shielding. Shielding, also known as casing, mandates about ten inches of steel and concrete to protect or "shield" underground aquifers. Shielding is an industry standard and is required in proper procedure. Resources include water, which is 90 % of fracturing fluids, with only .5 % of chemicals that are commonly found in cosmetics and household cleaning products (Energy From Shale 2013). To protect the groundwater supply, backflow preventers are installed. Through one-way water flow, contaminated water will be prevented from entering groundwater. The contaminated supply is held in storage tanks and monitored until drilling is completed and impoundments are properly removed. Recycling operations are also employed for future drilling usage.

With the streamlining of widespread knowledge and operations, stakeholders can evaluate their position in the fracking equation. Figure 2 illustrates the breadth of potential stakeholders relating to unconventional oil and gas drilling activities. Stakeholders influence the priorities and limits that can be set on shale plays. When mindful of both goals and regulations, rightful actions can be established leading to an economically promising progression. Recognizing how much is too much and whether these activities are safe will be continual questions to be asked in many production processes from production, regulation, and subsistence.

At an individual level, estimations of over 15.3 million Americans lived within a mile of a well that has been drilled since 2000. Those identified stakeholders have been greatly influenced by drilling practices, as 15.3 million is more than the population in Michigan or New York City (Quora 2014). But, this is one of many segmented groupings that have been influenced by fracking at any level, growing the individual influence rates over hundreds of millions.

Other stakeholders have significant involvement in unconventional oil and gas drilling activity. As discussed in several chapters in this volume, the job market considers current and future labor impacted by unconventional shale gas development. Those in financial markets include rights holders, lessors, investors, and analysts. Significant legal and regulatory measures and concerns exist; thus, regulators, legislatures, and related administrative and enforcement staffs are also stakeholders. These impacts ripple throughout the value chain of unconventional shale gas development. Lastly, there are a number of other groups who are stakeholders, including trade association and trade promotion groups, various nongovernmental organizations, local civic groups, and other special interest groups.

According to report commissioned by the US Energy Information, there are approximately ten times more oil and gas resources in shale outside the United States than inside the United States. Some exploration and development has begun in South Africa's Karoo Shale, the Vaca Muerta Shale in Argentina, as well as others in China, the United Kingdom, and Poland. The Eagle Ford Shale in South Texas is believed to extend beneath the international boundary into Mexico.

Fig. 2 Stakeholder analysis

The sustainability of the resources is as important as its value and job creation. Since America's first drilling in 1859, gas and oil drilling activity has grown to 1.1 million wells. It is important to find a balance between short-term drilling activity and sustainable development in the long run. How greatly can these activities affect the natural landscape or daily lives of consumers? The challenge moving forward is how these activities can provide benefit from a tremendous natural resource and not to leave all stakeholders at risk.

Will America prepare for a short-lived shale-rush or a sustainable energy-surging economy? The decision is for the current and the next generation to decide. We hope this volume will help inform readers about the reality, impacts, and potential of shale exploration.

Overview of This Volume

This book has 10 chapters. Following this Introduction, the chapters reveal various aspects of shale gas development through examples in different states, different areas of the states, different industry sectors, and supply chain.

Chapter 2 by Hefley and Seydor reveals the direct impact of shale gas exploration from a single well perspective. It examined the supply chain surrounding drilling and production of Marcellus Shale gas well in Southwestern Pennsylvania and evaluated the eight phases of the life cycle of a shale gas wellhead. It estimated the cost of developing a single well over seven million dollar in early 2011. With the development of technology and economy of scale, cost has decreased in the past several years.

After understanding the life cycle and impact of a single well, Chap. 3 by Wang and Stares selected Washington County, Pennsylvania, as the sample to evaluate the economic potential of shale gas development for a county. Washington County is a major area of Marcellus Shale development. The first well of Marcellus Shale was drilled in 2004 by Range Resources in this county. Many shale gas companies have either headquarters or offices in the county. Based on the economic and drilling information of 2011, it estimated that the economic potential of drilling and production activities of shale gas development could range between 8.9 % and 9.3 % of total output assuming a 15 % royalty rate and between 4.9 % and 5.3 % on employment. The exact impact highly depends on the localization level of these activities and the local spending rate of royalty earners.

Six counties of the Northeast Pennsylvania accounted for more than half of the shale gas wells drilled in Pennsylvania. They are Bradford, Lycoming, Sullivan, Susquehanna, Tioga, and Wyoming counties. Chapter 4 by Hardy and Kelsey explored the impacts of shale gas development in these counties. These are mainly rural counties without prior history of oil and gas development which is different than the experiences of counties in Southeast Pennsylvania, e.g., Washington County. From 2007 to 2011, the change of rents, royalties, patents, and copyrights income increased more than 500 % in this six-county region. During the same time, the change in Pennsylvania was 37 %. There was an increase of 166 % of employment in mining industry between 2007 and 2001 in the six-county region and 63 % in Pennsylvania.

Chapter 5 by Kelsey and Hardy showed that, in Pennsylvania, the impact of shale gas development was more pronounced on worker's compensation of related industries. There was positive, but modest, increase on employment. Although Marcellus Shale development accounted for a small portion of the workforce in Pennsylvania, it has been the major source of job growth in the past several years. Stronger impacts concentrated near counties that had active shale gas activities. There were minor impacts on non-drilling counties. Comparing to the size of capital investment in billions of dollars, the related impacts are relatively less dramatic.

In addition to Pennsylvania, another major shale gas region is Texas. Chapter 6 by Tunstall examined the impact of Eagle Ford Shale on the economy of its surrounding counties and the state of Texas. In 2012, the development at Eagle Ford generated over $61 billion in the economy and created 116,000 jobs. There has been substantial expansion of local infrastructure of rail and pipelines. This development helped some of the poorest local counties. The concern for future development is how to diversify the economy and have a sustainable growth after the shale gas boom.

Many of the regions of shale gas development are rural communities. This industrial activity could potentially alter the local culture and living in a dramatic way. Chapter 7 by Braiser et al. explored this issue with both quantitative and qualitative data. It evaluated various social indicators including housing, health care, education, crime, and residents' perceptions of their communities. Although quantitative data showed limited difference between regional and long-term trends, qualitative findings revealed substantial changes of perceptions of local residents about community changes and future outlook.

The above economic and social impacts are influenced not only by the practices of shale gas activities but also local and state regulations. The legal and regulatory frameworks at multiple levels play an important role in shaping the shale gas development. Chapter 8 by Murtazashvili discussed this issue by comparing Pennsylvania with New York. Pennsylvania reacted actively to shale gas development by creating formal legislation called Act 13. On the contrary, New York has a de facto ban on the development by having an extended reviewing and study process. The close proximity of these two states and similar resources in Marcellus Shale present a great example of study. After the evaluation, it showed that the regulatory response in Pennsylvania is "efficient" and that different responses of each state can be explained by features of politics, rather than geography, relative prices, or institutions. The findings also suggest decentralized governance by the states instead of the federal government has many benefits.

There are several industrial sectors that are directly impacted by shale gas development including electricity generation, transportation, and manufacturing. Chapter 9 by Krupnick et al. investigated these relationships. As a transitional cleaner fuel, natural gas started replacing coal as the power generating source in more and more power plants. Recent announcement of the Clean Power Plan by the Environmental Protection Agency (EPA) further promoted this trend. Although it is still in the very early stage, the use of natural gas in the transportation sector is projected to increase with more investments in refueling infrastructure and better natural gas vehicle technologies. Petrochemical and other manufacturing industries in the United States and abroad have responded to lower natural gas prices by investing in US-located manufacturing projects.

Energy is the foundation for a modern economy and society. Developing a new source of energy like shale gas is a complex process. It affects many different aspects. Chapter 10 by Lipscomb et al. summarizes these aspects including economy, sociodemography, environment, and regulation. It suggests the importance of having a balanced approach of development across these areas and pointed out topics worthy of future exploration.

The goal of this book is to provide insights to shale gas development and its impacts. By no means, it is comprehensive. The cases and studies in this volume are from the most active regions of shale gas exploration within the United States. We hope it will serve as a good starting point for communities, policy makers, the industry, and researchers to have informed, open-minded discussions.

References

American Petroleum Institute (2014) Hydraulic fracturing: Unlocking America's Natural Gas Resources. http://www.api.org/oil-and-natural-gas-overview/exploration-and-production/hydraulic-fracturing/~/media/Files/Oil-and-Natural-Gas/Hydraulic-Fracturing-primer/Hydraulic-Fracturing-Primer-2014-lowres.pdf

Barth J (2010) Unanswered questions about the economic impact of gas drilling in the Marcellus Shale: Don't Jump to Conclusions. JM Barth & Associates, Inc, Croton on Hudson. http://occainfo.org/documents/Economicpaper.pdf

Beaver W (2014) Environmental concerns in the Marcellus Shale. Bus Soc Rev 119(1):125–146

Black DA, McKinnish TG, Sanders SG (2005) Tight labor markets and the demand for education: evidence from the coal boom and bust. Ind Lab Relat Rev 59(1):3–16

Boudet H (2014) "Fracking" controversy and communication: using national survey data to understand public perceptions of hydraulic fracturing. Energy Policy 65:57–67

Considine TJ (2010) The economic impacts of the Marcellus Shale: implications for New York, Pennsylvania, and West Virginia. The American Petroleum Institute, Washington, DC. http://www.scribd.com/doc/34656839/The-Economic-Impacts-of-the-Marcellus-Shale-Implications-for-New-York-Pennsylvania-West-Virginia

Considine TJ, Watson R, Blumsack S (2010) The economic impacts of the Pennsylvania Marcellus Shale natural gas play: an update. The Pennsylvania State University, College of Earth and Mineral Sciences, Department of Energy and Mineral Engineering, State College

Crompton J (2013) Activists oppose drilling in area protesters criticize 'Piggish' industry in Lawrence County. Pittsburgh Post – Gazette, Jan 28, B.1

Earthworks (9 Aug 2011) "Hydraulic fracturing 101." Web. 22 Mar 2014

Energy from Shale "Groundwater Protection." Energy tomorrow, 2013. Web. 05 Apr 2014

Engelder T (Aug 2009) Marcellus 2008: report card on the breakout year for gas production in the Appalachian Basin. Fort Worth Basin Oil & Gas Magazine, 18–22. http://www.marcellus.psu.edu/resources/PDFs/marcellusengelder.pdf

Environmental Law Institute and Washington and Jefferson College (2014) Getting the boom without the bust: guiding Southwestern Pennsylvania through shale gas development. http://www.eli.org/sites/default/files/eli-pubs/getting-boom-final-paper-exec-summary-2014-07-28.pdf

FracFocus (20 July 2010) Hydraulic fracturing: the process. Accessed 28 Mar 2014

Gannon, Joyce (24 Jan 2010). Marcellus Shale group leader excited: talking with Kathryn Z. Klaber. Pittsburgh Post-Gazette (Pennsylvania), Sunday Two Star Edition, Business Section, C-1

Higginbotham A, Pellillo A, Gurley-Calvez T, Witt TS (2010) The economic impact of the natural gas industry and the Marcellus Shale Development in West Virginia in 2009. Bureau of Business and Economic Research, College of Business and Economics, West Virginia University

Hopey D (27 Feb 2011) Fracking 101: Jury is out on health, 'green' concerns. Pittsburgh Post-Gazette. http://shale.sites.post-gazette.com/index.php/background/fracking-101

IHS (2012) America's new energy future: the unconventional oil and gas revolution and the US economy, Volume 1: National Economic Contributions. http://marcelluscoalition.org/wp-content/uploads/2012/10/IHS_Americas-New-Energy-Future.pdf

Kay D (Apr 2011) The economic impact of Marcellus Shale gas drilling: what have we learned? What are the limitations? (Working paper series – a comprehensive economic impact analysis of natural gas extraction in the Marcellus Shale). Cornell University. http://www.greenchoices.cornell.edu/downloads/development/shale/Economic_Impact.pdf

KDKA (2014) Beaver County braces for Shale gas infrastructure. http://pittsburgh.cbslocal.com/2014/07/23/beaver-county-braces-for-shale-gas-infrastructure/

Kretschmann D (2014) Veggies and help, personal correspondence, July 15

Litvak A (16 Aug 2011) Marcellus Shale production jumps in southwest Pennsylvania. Pittsburgh Business Times. http://www.bizjournals.com/pittsburgh/blog/energy/2011/08/large-spike-in-marcellus-production.html

MIT Energy Initiative (2011) The future of natural gas: an interdisciplinary MIT Study. Massachusetts Institute of Technology, Cambridge, MA. http://web.mit.edu/mitei/research/studies/documents/natural-gas-2011/NaturalGas_Report.pdf

Mufson S (2014) How two small New York towns have shaken up the national fight over fracking. Washington Post, July 2. http://www.washingtonpost.com/business/economy/how-two-small-new-york-towns-have-shaken-up-the-national-fight-over-fracking/2014/07/02/fe9c728a-012b-11e4-8fd0-3a663dfa68ac_story.html

Natural Gas Weekly (19 July 2010) Fracturing ban not likely, but compliance Costs likely to rise, 2–3

Novak S (2014) Marcellus to drive $70 billion in infrastructure development. Pittsburgh Business Times. Energy Inc, 18 Mar 2014. Web. 1 Apr 2014

Olson L (19 Aug 2011) State gas production spikes; 6-month figures show 60 % increase for Marcellus Shale wells. Pittsburgh Post-Gazette 85(19): A-1, A-7

The Perryman Group (2008) Drilling for dollars: an assessment of the ongoing and expanding economic impact of activity in the Barnett Shale on Fort Worth and the surrounding area. Waco, The Perryman Group. http://www.bseec.org/sites/all/pdf/report.pdf

Public Utility Commission (2014) Act 13/Impact fees. https://www.act13-reporting.puc.pa.gov/Modules/PublicReporting/Overview.aspx

Pursell D (2010) Natural gas supply study update, 15 Oct 2010. Tudor, Pickering, Holt & Co. Securities, Inc, Houston. http://tudor.na.bdvision.ipreo.com/NSightWeb_v2.00/Handlers/Document.ashx?i=bb694381e0774244ab6eb028b8c73910

Quora (2014) How many people are affected by fracking? Forbes. Forbes Magazine, 24 Mar 2014. Web. 28 Mar 2014

Sample V, Price W (2011) Assessing the environmental effects of Marcellus Shale gas development: The State of Science. Pinchot Institute for Conservation, Washington, DC. http://www.pinchot.org/uploads/download?fileId=963

Smith G (2013) U.S. to be top oil producer by 2015 on Shale, IEA Says. http://www.bloomberg.com/news/2013-11-12/u-s-nears-energy-independence-by-2035-on-shale-boom-iea-says.html

Taylor K, Kaplan T (2014) New York towns can prohibit fracking, State's top court rules. The New York Times, 30 June. http://www.nytimes.com/2014/07/01/nyregion/towns-may-ban-fracking-new-york-state-high-court-rules.html

Ubinger JW, Walliser JJ., Hall C, Oltmanns R (2010) Developing the Marcellus Shale: environmental policy and planning recommendations for the development of the Marcellus Shale Play in Pennsylvania. Harrisburg: Pennsylvania Environmental Council. http://www.pecpa.org/sites/pecpa.org/files/downloads/Developing_the_Marcellus_Shale_1.pdf

U. S. Department of Energy, Office of Fossil Energy, National Energy Technology Laboratory (2009) Modern shale gas development in the United States: a primer. National Energy Technology Laboratory, Strategic Center for Natural Gas and Oil, Morgantown. Available at http://www.dep.state.pa.us/dep/deputate/minres/oilgas/US_Dept_Energy_Report_Shale_Gas_Primer_2009.pdf

Williams S (2014) Britain to award licenses for shale-gas exploration. Wall Street J CCLXIV(23), July 28, B2

Direct Economic Impact of the Value Chain of a Marcellus Shale Gas Well

William E. Hefley and Shaun M. Seydor

Abstract This chapter examines the direct economic impact of a Marcellus Shale well located in Southwestern Pennsylvania. This study is an assessment of the economic impacts emphasizing the direct economic impact, rather than just focusing on the perceived benefits and impacts affecting the region. Our analysis is based on extensive field research, including a site visit and interviews with industry participants. From this field research, we determined that the direct costs of bringing a Marcellus Shale well to production are in excess of seven million dollars.

Economic Impact of a Marcellus Shale Well

The focus of this chapter is on the direct economic impact of Marcellus Shale development. There can be several types of economic impact from a particular economic activity. These can be categorized as direct effects, indirect effects, and induced effects. This study examines the direct effects of a single Marcellus Shale well, developed using horizontal drilling and hydraulic fracturing, in Southwestern Pennsylvania. By using a single well as a standard unit of measure, this study can help to better understand the Marcellus Shale. This project sought to quantify the "business" factors of a single Marcellus Shale well value chain, by understanding the direct spending in preparing, drilling, fracking, and moving into production a single Marcellus Shale well site.

Many of the existing economic impact studies are based on input–output models (Miller and Blair 2009; U. S. Department of Commerce 1997). Barth (2010) makes the argument that the labor flows in Marcellus plays may not match the underlying assumptions

W.E. Hefley (✉)
Joseph M. Katz Graduate School of Business and College of Business Administration, University of Pittsburgh, Mervis Hall, Pittsburgh, PA 15260, USA
e-mail: wehefley@katz.pitt.edu

S.M. Seydor
Institute for Entrepreneurial Excellence, Innovation Institute, University of Pittsburgh, Pittsburgh, PA 15260, USA
e-mail: iee@innovation.pitt.edu

W.E. Hefley, Y. Wang (eds.), *Economics of Unconventional Shale Gas Development*, Natural Resource Management and Policy 45,
DOI 10.1007/978-3-319-11499-6_2

in the input–output models. Thus, to address this concern as well as Crompton's (1995) caveat that studies should explicitly account for costs, the current study focuses on the direct economic impacts of Marcellus Shale drilling in a single Marcellus Shale well.[1]

While the direct economic impacts in this chapter were estimated using costs for exploration and production companies that extensively make use of service providers to provide much of the equipment and labor, our earlier study (Hefley et al. 2011) cross-validated these costs by examining the cost structure of a firm that emphasizes low costs and vertical integration.

Phases of the Lifecycle of a Marcellus Shale Wellhead

The development of a wellhead typically progresses through a lifecycle consisting of multiple phases, with each phase composed of multiple steps. The steps within each phase could vary across sites, depending on factors, such as the current drilling or leasing status of the site and its geography. Lifecycle phases of a typical wellhead are:

- Phase 1 Mineral Leasing/Acquisition and Permitting
- Phase 2 Site Construction
- Phase 3 Drilling
- Phase 4 Hydraulic Fracturing
- Phase 5 Completion
- Phase 6 Production
- Phase 7 Workovers
- Phase 8 Plugging and Abandonment/Reclamation

Figure 1 provides a visual depiction of these phases and key steps. An enormous amount and variety of inputs from various sources come together for one drilling site. The value chain begins with site preparation and continues all the way through postproduction. The site needs to be leveled, with proper entrance and exit roads for the equipment. Then, all the actual drilling equipment is put into place, which may require the rental of the equipment, with truckloads transporting the equipment to the site. Before drilling, a sustainment infrastructure needs to be put in place. This includes generators to provide power to the entire site, which use non-road diesel that needs to be transported on-site, and may include living quarters for the drilling workers. Security measures may be put into place. All water used throughout the process either needs to be piped or trucked on-site. Then, when the drilling starts, all of the ingredients for the lubricating "mud" need to be bought and transported, including water, salt, and a mix of chemicals. Then, the mud is processed and most of it is recycled and drilling chips separated and trucked away. After the vertical drilling is complete, concrete filler is put in place to keep the integrity of the hole, protecting both the well itself and the environment that it traverses. Then, the

[1] Crompton (2006) concludes that the "motives of a study's sponsor invariably dictate the study's outcome." To overcome these common limitations of economic impact studies, this study was not sponsored or funded by exploration and production firms in the Marcellus Shale industry.

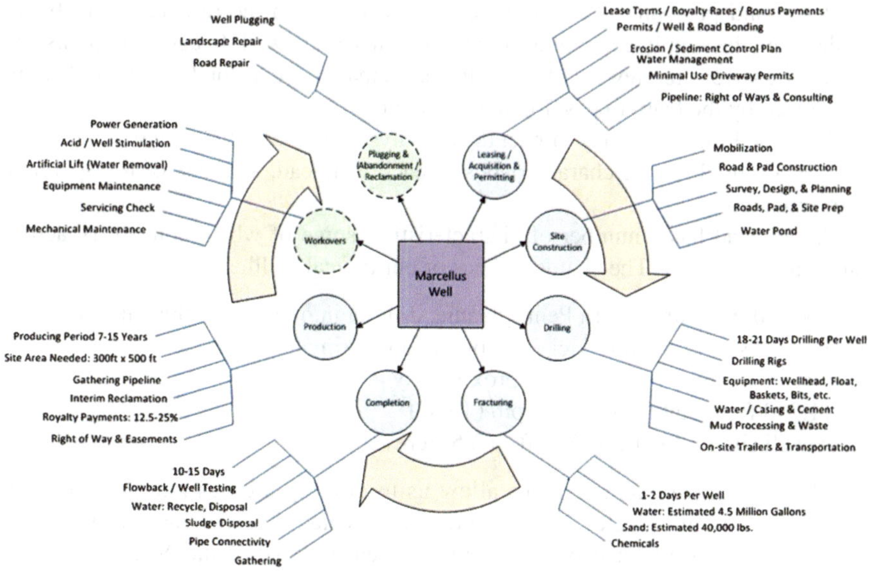

Fig. 1 Phases and key steps in developing a Marcellus Shale well site

horizontal drilling process starts, which also requires the lubricating "mud." When complete, the horizontal section gets the concrete as well. Next in the value chain is the shale fracturing process. This process requires the charges that will be put underground as well as the fracturing fluid which consists of water, sand, and another mix of chemicals and additives. The outflow of fracturing fluid also needs to be either held temporarily on-site and transported off-site or immediately transported. After this process, the equipment is removed and the piping infrastructure is put into place along with a permanent wellhead or "Christmas tree."

This chapter addresses the direct economic effects of Phases 1 through 6 of a Marcellus Shale well. Phase 7, occurring throughout the working life of a producing well, and Phase 8, which occurs at the end of the life of a well site, are not included in our analysis. Given the expected productive life of a well spanning over many years, these costs will indeed have continuing economic benefit to the region, but are not addressed in this study. Each of these phases in developing a producing Marcellus Shale horizontal well is briefly described in the following sections. More details on each step can be found in Horizontal Drilling Animation (Louisiana Oil and Gas Association 2008) or other reports (U. S. Department of Energy 2009).

Data Collection

A key source of developing the research team's understanding of the Marcellus drilling and fracking process was a site visit, made by the entire research project team, to an in-process well site in Washington County, Pennsylvania. Access to the well site and personnel were provided by EQT to help the research team better understand the supply chain of a single Marcellus Shale well. In addition to this site visit, extensive interviews with industry participants and secondary research were conducted by the team to develop their cost model.

Multiple limitations may impact these analyses. These include the uniqueness of each well, the differing characteristics of each wellhead, and lack of transparency into actual costs.

A wellhead has a number of characteristics, some of which could vary across sites and geography. The characteristics of our typical wellhead are:

- Located in Southwestern Pennsylvania, drilling into the Marcellus Shale deposit
- Vertical shaft drilled to kick-off point at approximately 6,000 feet
- Single horizontal lateral, of approximately 4,000 feet
- 11,000 foot total measured depth (TMD)
- A well site of 300 ft. by 500 ft. = 3.5 Acres

These assumed characteristics allow us to develop a cost model of a typical wellhead and are reflected in the value chain, which describe each phase of the lifecycle and the direct economic impact of each phase of the Marcellus Shale extraction lifecycle.

This analysis was developed in early 2011, before the slowdown in drilling activity, and reflects costs as of that time. Numerous data sources were used including laws and regulations, public records, published literature, observations and interviews from site visits to Marcellus wellhead, and numerous telephone and email interviews with industry participants.

Value Chain of a Marcellus Shale Wellhead

Building on the lifecycle presented in Fig. 1, this section summarizes the value chain of a Marcellus Shale wellhead by examining the total spend associated with a typical wellhead in Southwestern Pennsylvania. It follows a general lifecycle flow, detailing specific steps within the lifecycle and their costs, to develop the value chain of a typical Marcellus Shale wellhead.

Phase 1 Mineral Leasing/Acquisition and Permitting

When analyzing the total cost of drilling a gas well, two preliminary steps must be considered: mineral leasing and acquisition and permitting. These steps are critical to the establishment of a well and can contribute significantly to overall cost.

Exploration and production companies, or landmen acting for them, must approach and negotiate with landowners for mineral rights leasing (see Table 1 below for examples of Standard Terms and Conditions). This process will often start with

Table 1 Standard terms and conditions

Term	Notes	Term	Notes
Term	Primary/Secondary	*Wells*	Disposal and injection
Royalties	–	*Pooling*	–
Delay rentals	Paid-up vs. Annual	*Pugh clause*	Vertical and horizontal depth
Shut-in	Price, duration	*Depth limitation*	Marcellus or other strata
Force majeure	–	*Taxes*	Severance, ad valorem
Surface use/ non-surface	Limited use, equipment limitations, location, Road Widths, pipelines	*Surrender and termination*	Right to surrender, equipment removal, termination/survival of easements, recording
Surface damages	–	*Implied duties*	Protect from drainage, etc.
Easements	Pipelines, access roads, etc.	*Audit*	–
Water quality	Pre-drill testing, replacement	*Dispute resolution*	ADR, jurisdiction
Water use	Ponds, streams, wells, etc.	*Other*	Needs of lessee/Lessor
Gas storage	–	–	–

the largest tract of land, moving on until sufficient rights are acquired for effective production. This study assumes that 320 acres is the minimum acreage to permit, with 640 acres (1 square mile) being the minimum optimal size. Adjacent properties may also be placed under license, as surface/non-surface leases allow placement of the pad site location on property or only the access to minerals underneath.

Second, the permitting stage requires the satisfactory filing/obtaining of state and local permits and posting of necessary bonds to allow for site preparation to begin.

Leasing/Acquisition

The acquisition of mineral rights and development of a proposed unit is the first step in the development of the Marcellus Shale drilling process. The leasing and acquisition stage begins with the assumption that adequate and appropriate land has been identified. Geological exploration and its associated costs are therefore excluded from our analysis in this study.

Landowners, also known as lessors, will lease their respective mineral rights, specifically the oil and gas, underneath the property of which they have ownership. The primary benefits that can be recognized from the signing of a lease are in the form of a signing bonus, also known as a paid-up lease, and royalty rates.

Landmen have the principal responsibility of approaching and acquiring landowner's mineral rights by leasing the parcel with a number of negotiable terms to be considered as a binding agreement. The landmen typically represent an operating company, whereas the operating company is known as the lessee. The landmen must establish a unit that contains a minimum of 640 acres (1 square mile) of land that contains adjacent parcels in order to reduce the amount of petitioning rights to gain the privilege of drilling commencement. The analysis that covers the cost of acquisition will be based on this amount of acreage. In order to determine the actual mineral interest of property, title checks are done to determine that the correct parties have been signed and are able to release their rights for a specified time period.

The landowner's greatest incentives come from a few different contingencies of the lease. The most important of the leasing conditions, which have been mentioned above, are that of the signing bonus and the royalty rates. The signing bonus is the "short-term" amount that entices owners to sign the rights of their land to an operator for a certain time period. The signing bonus is negotiated separately from the royalty rate, and in most instances, is the only driving force for the parcel owner to sign so that an instant profit can be seen from the arrangement. There is a relatively high possibility that a leased property will not see a completed well due to the location or inability to establish a unit, or for other active wells or mineral reservations in the area. For this reason, the signing bonus becomes the most important factor in the negotiations due to the possibility of it being the only source of revenue that will be seen. The average signing bonus is found to be $2,700/acre (www.pagaslease.com). Using this estimate, the overall cost of signing bonus (640 acre unit) is $1,728,000. This is based upon the fact that all landowners that have been pooled into the unit have been offered the same amounts. This amount, as well as others, is highly variable, and the breakdown of the average scenario can be found in Table 2. The average lease is estimated to be a 50-year primary term. In addition, the operating company may have the ability to extend the lease for an additional 5-year term, at which time the property owner will receive the signing bonus again. This will then double the amount of cost for each lease that needs to be renewed within the unit to be

Table 2 Average costs of land acquisition

Acquisition/Leasing	Based off 640 acre site	1 year = 250 days			
Parcels in pooled unit=	50				
Labor Costs	Avg. Time (days per padsite)	Rate (avg day rate)		Total	
Leasing					
Landman	375	$	300.00	$	112,500
Title research					
Abstract (*per unit*)	10	$	275.00	$	137,500
Curative (*per unit*)	25	$	275.00	$	6,875
Subtotal (subject to # of parcels in unit)				*$*	*256,875*
Leasing Costs (paid-up lease)	*Avg. Cost*	*Amount*			*Total*
Signing bonus/acre	*2700*		640	$	1,728,000
Bonus/padsite location	10000		1	$	10,000
Shut-in (*typically not paid*)	10		640	$	6,400
Lease filing at courthouse (*per parcel*)	78.5		50	$	3,925
Subtotal(subject to # of parcels in unit)				*$*	*1,934,250*
Total				*$*	*2,191,125*

established. The reason for this is that if the unit has not been developed within the first 5 years, then the lease extension will grant them the ability, if it so chooses to, to complete the unit and drill during the extended lease terms.

The royalty rate is a percentage of the produced amount based on the completed well's output. For example, in a 640 acre unit (1 square mile), suppose a landowner owns 320 acres of the established unit. If the royalty rate is agreed to be 1/8, or 12.5 %, then the profitability that the landowner would recognize would be 12.5 % of ½ of the total amount produced from the wellhead. The remaining 320 acres will be dispersed accordingly with regard to the remaining property owner's respective royalty rates and acreage in relation to the total sum of the unit.

Another area of concern is the type of lease that is signed. Landowners have the ability to lease a surface or non-surface lease. The surface lease allows the operating company the access to have the pad site location to be on their property. The non-surface lease allows the operator to only drill underneath the property to access the oil and gas. This is achieved through the pad site location being within the capable distance of a drilled lateral. Mainly, non-surface leases are paid a lower amount, due to the restricted access, as well as many times being too small of an area to be considered for a pad site location. The parcel owners that do sign a surface lease, typically greater than 5 acres for a pad site to be located, are often times given additional payments based on the pad site being located on their property. Average amounts of bonuses are estimated to be $10,000 if their property is selected to be utilized for the drilling location (Title Abstractor "C", personal communication, April 12, 2011).

Shut-in fees are the last aspect of the cost of leasing parcels. Although the shut-in fees typically are not of concern, and for this reason have not been estimated to be a cost of acquisition, they are able to generate a cost to the operator. Shut-in fees are a predetermined amount that is paid to the landowner in the event that the well is stopped from producing due to any number of reasons on the operator's behalf. In the event that production is available and is stopped, the landowner will receive an amount, which is typically minimal, to be paid on a daily, monthly, or annual basis. Shut-in fees are typically not of concern, as the Marcellus Shale operating companies fully recognize the need to produce and sell the natural gas in order to profit from its capital requirements.

In order to establish a unit of land to be capable of drilling, the landmen approach and negotiate the stipulations of the lease on a case-by-case scenario. The landmen are often times hired by the operating companies as independent contractors that are paid a day rate for their services. The average amount of time that it takes for a single landman to develop a unit for drilling purposes is estimated to be around one and a half years, or 375 working days (Landman, personal communication, April 17, 2011). At an average day rate of $300/day, the associated leasing labor is estimated to be $112,500. Again, these numbers are an average, and labor rates can vary from $150/day and up to $450/day, not including per diems and paid mileage (Landman, personal communication, April 17, 2011). The time required is also highly variable, with a best-case scenario of 9 months and a worst-case of 5 years. The amount of surrounding acreage and the willingness of the mineral owners are highly correlated to this cost. The number of parcels that are involved in the pooling of a unit fluctuates from a minimal number, such as 5, to as many as 500, depending on acreage sizes of surrounding property owners. With larger numbers of parcels that need to be approached and negotiated with, this process can become highly involved and can take years to complete. For the purpose of generating an average number of parcels, 50 properties are considered as a benchmark (Landman, personal communication, April 17, 2011).

After a lease is signed, the determination of the mineral interest is researched. The parcel of land is researched initially by the landman to conduct a due-diligence research on the property. This entails running the title back to the approximate time between 1850 and 1880. The date that it is targeted to be researched to is determined by the initial drilling and exploration that Pennsylvania has been exposed to. The first wells were drilled around this time and can affect the ownership of parcels to current date. In order to determine the interest, a few steps are taken at the county courthouse's recorder of deeds office in which the parcel is located.

After the preliminary title check is done and approved, the lease is taken to the county courthouse and filed to be recorded into the system of publicly available information. This is estimated to be $78.50/lease. Based on a unit size of 50 parcels, the amount is estimated to be just under $4,000/unit. The recording of the lease document allows other companies to realize the ownership of that lease for further research and unitization pursuance.

The estimated average time to conduct an abstracted title averages 10 working days, with a best-case scenario of 5 days and worst-case scenario of 6 months (Title Abstractor "B", personal communication, April 4, 2011). The title researchers are typically sourced and paid as independent contractors. The typical day rate is averaged to the amount of $275/day, with variances of $150–$400/day seen (Title Abstractor "A", personal communication, April 4, 2011). This is highly variable for the associated contractor's experience and paying company involved.

The amount of parcels that are in the developed unit is what consumes most of the abstracting costs. With the number of parcels averaging 50 in a given unit, the average day rate of $275 is estimated to have an overall cost of $137,500/unit (Title Abstractor "C", personal communication, April 12, 2011). This number is extremely susceptible to fluctuation based on the complexity of the title and the number of parcels in a unit. The best-case scenario is based on a 5 parcel unit at $150/day rate, with each parcel requiring 5 days of working time, equaling $3,750 of labor. The worst case, on the other end of the spectrum, is based on 250 parcels being evaluated at $400/day and 6 months working time. This translates to an astronomical increase of $12 M/unit. At this rate, the operating company would not benefit from a profitable situation from the well's production.

The last part of the acquisition cost involves the title research and acquisition known as curative title and development. The average amount of curative work needed to be done for each unit is estimated to be 25 working days (Title Agent "A", personal communication, April 14, 2011). The variability of this may differ from 10 working days (2 weeks) up to 120 days (6 months). The average labor rate is based on an independent contractor day rate as well, with average rates of $275/day being an estimated average (Title Agent "B", personal communication, April 19, 2011). The rate fluctuates to include variances of the same amounts of the abstracting department, being $150–400/day. Total average cost per unit for curative research and procurement is estimated to be $6,875.

In the end, the complete cost of the leasing and associated labor costs generated from landman and title research is estimated to be approximately $2.2 million/640 acre unit. The amount of variability depending on numerous conditions and circumstances reflects a best-case scenario of approximately $100,000 and a worst-case scenario of approximately $20.7 million. The amount of complexity and parcel acreage, as well as landowner's willingness to lease, can prove to generate numbers at any point within these scenarios.

Table 2 depicts the average scenario for overall costs of land acquisition. Hefley et al. (2011) provide more details of sensitivity analysis around costs of bringing the well to production.

Permitting

Total permit application costs for Marcellus Shale wells include three components: permit application, abandon well surcharge, and orphan well surcharge. When drilling a gas well in Pennsylvania, the well operator must obtain a well permit from the Department of Environmental Protection (DEP). The permit application must show the location of the well, proximity to coal seams, and distances from surface waters and water supplies. Technical staff in DEP's Regional Offices reviews the permit application to determine whether the proposed well would cause environmental impacts and conflict with coal mine operations.

To address additional environmental considerations associated with development of shale, the DEP developed an addendum specifically for shale gas well development. The DEP expends considerable staff resources to review the additional information in the Marcellus Shale Addendum because the review includes several water quality and quantity issues not normally associated with gas well permit application reviews.

Effective April 18, 2009, the application fee for well permits for shale natural gas wells follows a sliding scale based on wellbore length and type. Any application received on or after April 18, 2009, must include the new application fee in addition to the surcharge fees for abandoned wells and orphan wells.

The permit fee is based on the anticipated total length of the wellbore in feet, which is the total measured depth (TMD) for horizontal wells. If the well is drilled longer than what was applied for in the application, the applicant will be required to pay the difference between the amount paid on the original application and 10 % on the amount required by the completed wellbore length. The surcharge can be avoided by amending the original permit and paying an additional permit fee. A refund is not issued for under-drilling the length of a permit.

The permit fees for the gas wells were established to cover program costs including hiring additional staff in Meadville, Pittsburgh, and Williamsport to process permits and better monitor drilling activities statewide. 25 PA Code § 78.19 Permit application fee schedule defines the fee structure for obtaining the required state permit to drill a well. Using our assumed well with a total measured depth of 11,000 feet, the permit cost is $3,050. Using a more typical well site with vertical depth of 8,000 feet and 3 horizontal bores of 4,500 feet, the total permit cost is $5,150.

Table 3 Permit fees and bonds required

25 PA Code § 78.19 State permit (drilling) fees		
Permit fee	$4,900	
Orphan well surcharge	$200	
Abandoned well surcharge	$50	$5,150
25 PA Code § 78.310 Well bonding		$2,500
25 PA Code § 102.6 Erosion, sediment control plan		$1,900
25 PA Code § 91.22 Water management		$500
67 Pa. Code § 441.4 Minimal use driveway permits		$25
Total		$10,075
Plus		
67 PA Code § 189.4 Road bond for overweight vehicles		$12,500 per mile

In accordance with the Pennsylvania Oil and Gas Act, Marcellus Shale wells are "subject to orphan and abandoned well surcharges" of $200 and $50 per well, respectively. These surcharges are in addition to the gas well permit fees and will be paid into the Orphan Well Plugging and Abandoned Well Plugging Funds, as shown in Table 3.

Gas wells drilled in Pennsylvania after April 17, 1985 are required to be bonded, according to 25 PA Code § 78.310 Well Bonding. The bond is a financial incentive to ensure that the operator will perform the drilling operations, address any water supply problems the drilling activity may cause, reclaim the well site, and properly plug the well at the end of the wells useful life in accordance with their permit. The bond permit for a single well is $2,500; a blanket bond to cover any number of wells is $25,000.

25 PA Code § 102.6 addresses fees for erosion and sediment control. Fees are set at $1,500 plus $100 per acre disturbed. Thus, assuming a well site of 300 ft. by 500 ft., or 3.5 acres, results in a fee cost equaling $1,900.

The code requires all projects that disturb earth in the state to develop an erosion and sediment pollution control plan and implement best management practices for the control of sediment pollution during drilling. The Erosion and Sedimentation Control Program ensures that proper site development practices are employed for land development.

Generally, two different kinds of water use have to be differentiated in the context of a Marcellus Shale well. First, there is water for drilling. Secondly, the fracking process requires a significant amount of water. Water is obtained from several sources, including surface water locations such as rivers, streams and large lakes, and groundwater wells. All of these sources must be approved by the Pennsylvania Department of Environmental Protection (PADEP) and Susquehanna River Basin Commission (SRBC), if applicable. Gas companies also have water sharing agreements with other operators to reduce the industry impact.

Since August 14, 2008, gas companies are required by the Susquehanna River Basin Commission to seek permission to withdraw or use water to establish wells in the Marcellus Shale in the Susquehanna watershed. Without approval by the commission, gas companies are not allowed to start gas well construction, drilling, or hydrofracturing (Abdalla and Drohan 2010).

Fees for water quality management permits are addressed in 25 PA Code § 91.22. A Marcellus Shale gas well permit application includes an addendum for a water management plan that the operator must also submit to the DEP. The addendum is required due to the volume of water that is used in the hydraulic fracturing of the shale. The permit review evaluates the water intake information during the fracking process, in addition to the management, treatment, and discharge of the wastewater. The review of the water management plan requires additional DEP staff time because it requires staff to evaluate water intake information associated with the hydraulic fracturing of the shale, including review of the management, treatment, and discharge of the wastewater. The cost of this additional permit is $500.

When a well site is larger than five acres, a storm water management permit must be obtained. This "disturbed area" includes well sites, associated roads, pipelines, and storage areas to be constructed. The affected surface landowner and coalmine operator have the opportunity to file an objection about the location of the well. If DEP's permitting staff finds that no adverse impacts would result, the operator will receive a permit to drill the well.

67 PA Code § 189.4 establishes a road bond for overweight vehicles, resulting in bond charges of $12,500 per road mile. Road bonds for overweight vehicles can be provided in several forms: performance bonds issued by an insurance company, certified check, cashier's check, irrevocable letter of credit, or self-bonding if qualifications are met. Amounts are based on the type of roadway traveled and the maintenance required to repair the road due to the overweight vehicle. They are set in regulation at $6,000/mile for unpaved roadways, $12,500/mile for paved roadways, and $50,000/mile for paved roadways that are reverted back to unpaved conditions. A hauler traveling over numerous posted roads under the control of one owner can provide $10,000 security for each owner.

67 Pa. Code § 441.4 establishes fees for minimal use driveway permits of $25.00. The ability of a driveway to safely and efficiently function as an integral component of a highway system requires that its design and construction be based on the amount and type of traffic that it is expected to serve and the type and character of roadway which it accesses. Driveways are categorized into four classifications, based on the amount of traffic they are expected to serve. For purposes of a gas well, the minimum use driveway is applicable. Not more than 25 vehicles per day can use a minimum use driveway.

Phase 2 Site Construction

The second phase, Site Construction, involves the design and layout of the well site for the construction of the road and pad, or "staking the well." The steps involved in this activity include, among other things, survey, site design and layout, water planning (i.e., planning for water ponds, water supplies via trucks, or pipeline), construction of access roads, road and pad construction (i.e., staking the well), placement of on site trailers, construction of water storage or pits, and erosion control.

The process for site construction begins when companies are invited to bid on the site building project. Anywhere from 3 to 20 companies may be bidding on a site depending on the area that will be built upon and the exploration and production company building the site. The company who will be drilling the well gives each of the bidders a site plan with layout size and location, and the bidder is also able to go to the site to view it.

The first step in the construction process for the company awarded the bid is to call the utilities for the "One Call." This is where the utility companies such as data, gas, and water companies come out to the site and mark the utility infrastructure in place with flags so that the site construction does not damage any of the current lines in place.

The second step to the process is to determine what type of erosion control needs to be put in place. Erosion control is put into place to protect creeks, streams, and highways from damage, which can be caused if too much sediment washes off of a site while the soil is being disturbed by construction. The Department of Environmental Protection determines what type of silt protection must be put in place. This step of the process can cost from $10,000 to $20.000 provided a silt fence secured with wooden stakes or a silt sock can be used.

Once the erosion control plan is in place, the roads can be constructed to mobilize the equipment needed to construct a site. Costs can vary greatly by road length and type; however, average road construction for a site in Pennsylvania is from $10,000 to $20,000.

Mobilization is the process of moving the equipment to the work site and cost on average $10,000–$20,000. During mobilization, equipments such as dozers, backhoes, tractors, blades, rollers, and haul trucks are moved to the site. This construction equipment is moved to the site by a heavy haul company. This equipment will be used to level the site and create the foundation for the pad which is primarily constructed of stone.

Once the equipment is on-site, the site must be stripped and grubbed. The stripping process is when any trees on the land are cut down. Any tress over 6 in. in diameter thick can be sold by the land owner to be used for lumber. Trees under 6 in. are disposed of and can be used for wood chips. Grubbing the land removes any brush and tree stumps. Stripping, also known as timber removal, is often contracted out to a third party. Depending on how heavy the tree and scrub cover is, this process can cost from $0 for a natural field to about $45,000 for a more densely treed area.

After the area has been stripped and grubbed, the location is leveled. This process begins with the topsoil being stripped and reserved. The top soil needs to be saved to be spread back out over the area during the interim reclamation so that the area is able to be seeded. The process of leveling the location is similar to leveling a location for any type of build out. The area must be dug out or filled in to create a level lot. The location also has a 40 inch berm to contain any type of water or fluid spill. The berm protects the surrounding area from contamination should any fluids be spilled. Leveling a location in Pennsylvania costs on average $125,000–$300,000, as the landscape is primarily marked by hills making leveling necessary.

Table 4 Average costs associated with site construction

Step	Cost
One call	–
Erosion control	15,000
Roads	15,000
Mobilization	15,000
Strip and grub	23,000
Level location	213,000
Pond and liner*	70,000
Rock	15,000
Seeding and matting	35,000
Total	*400,000*

*Based on a $40,000 liner

At this point, a frack pond would be built if one was needed. The average cost for a frack pond is $60,000–$80,000.

After the earth work for a location has been completed, the pad is then constructed of rock. The base of the pad is 8–12 inches thick and constructed of a coarse aggregate. On top of this layer is 3–4 inches. of aggregate referred to as crush and run which is a finer aggregate material with smaller particles in it. When the crush and run is rolled using a smooth barreled roller, it appears similar to a parking lot. On average, a site requires $10,000–$20,000 worth of rock. The average price of rock is $25–$30 a ton, with some variation for proximity to a quarry.

Once the pad is constructed, the final stage in building the site is to seed the slopes on the outer edges of the site, as well as the berm. Seeding and netting (or matting) is done to help reestablish vegetation to prevent soil erosion. Proper erosion control is in place when 75 % vegetation is achieved. This process can cost from $20,000 to $50,000 per site. This is an important process in protecting the areas around the site from erosion damage.

When the site construction is complete, the equipment is mobilized off of the site and the next steps of the process can begin. Table 4 summarizes the costs associated with site construction.

Phase 3 Drilling

The drilling phase may take 23–35 days per well, including five days for mobilization and 18–21 days for drilling itself. This phase requires myriad pieces of equipment supporting drilling rigs, power generation, processing and disposal of liquid and solid waste (both chips from drilling operation and drilling mud returned with the chips), and the wellhead equipment and the Bottom Hole Assembly (BHA).

While this study focuses on a single well on a pad in a site, it is possible to place up to six wells per drilling pad, with each well having one or more horizontal laterals.

A Marcellus Shale natural gas well drilling operation can be broken down into two distinct phases. During the first phase of the process, a vertical wellbore is drilled down to a point just above the Marcellus Shale, and casing is placed into the wellbore. The casing not only protects the integrity of the wellbore from collapse, but more importantly it protects any water aquifers through which the wellbore passes. The second phase of drilling a Marcellus Shale well utilizes some of the newest technologies available to the industry. Drilling contractors will use down-hole motors and electromagnetic survey equipment to steer the drill bit in any direction while drilling a wellbore reaching thousands of feet through a seem of Marcellus Shale that sometimes is less than 20 feet thick. The horizontal portion of the well allows for the wellbore to have much more surface area; resulting in much greater amounts of gas that can be extracted. The benefit of this drilling technique is that a single horizontal well can produce the same amount of gas as six to ten vertical wells. Although there are various components of each section that are found in both the horizontal and the vertical stages, the costs of these are distinct to each stage of the drilling process.

The total cost of drilling is contingent upon the final depth and length of the wellbore. The Marcellus Shale formation lies approximately 7,000 feet below surface in the Southwestern Pennsylvania area of the Appalachian Basin. Once the vertical portion of the wellbore is drilled to a depth just above the Marcellus Shale (approximately 6,000 feet), the section of the wellbore referred to as the "curve" begins. This curve section will generally take 1,000 vertical feet to drill. The depth, at which the curve lands and becomes horizontal, or parallel with the surface, is commonly referred to as the total vertical depth, or "TVD." The horizontal portion of the well will be drilled approximately 4,000 feet straight out from the bottom of the curve and running within the Marcellus Shale the entire way. The result is a wellbore approximately 11,000 feet in total measured depth, or as it is commonly known in the industry, TMD.

Due to the high cost involved, most production companies do not own and operate their own drilling rigs. Instead, a production company will contract this work out to companies that specialize in the drilling process.

It is common for two different drilling rigs to be utilized during the drilling of a single Marcellus Shale well. A smaller rig that drills in a manner referred to as "air drilling" first drills the vertical part of the wellbore leading directly down to just above the Marcellus formation. Air drilling rigs pump high volumes of air down through the drill bit and use the air to carry the cuttings back to surface. A second and most times much larger rig is then moved in to drill the horizontal phase of the wellbore. This larger rig uses water-based or oil-based drilling fluid, commonly referred to as "drilling mud" to circulate the cuttings back to surface during the drilling operation. It is necessary to use a fluid drilling rig for horizontal phase of the wellbore due to the fragile nature of the Marcellus Shale. The fluid is noncompressible; therefore, it holds the wellbore open for around the drill pipe throughout drilling operations until casing can be ran in the wellbore. The heavy weight (usually between 12 and 14 lbs per gallon) of the drilling fluid also helps hold down any unexpected gas pressure that may be drilled into (generally referred to as a "kick").

For a typical well site, the total cost of the horizontal drilling rig rental, along with the cost of labor, averages $225,500 for a well that takes between 25 and 30 days to drill. Overseeing the operation and logistics of the drilling operation is a Drill Site Manager, whose fee averages $25,500. In addition to these costs, the production company must pay for the mobilization and assembly of the drilling rigs, with an average cost of $32,250.

During each drilling phase, the drilling rig is contained within a special containment area encased in pit liners. These liners can cost approximately $24,000 per site and are only in place to prevent contamination to the soil if there are any unplanned releases of fluids from either the horizontal or vertical drilling rigs.

Additional costs of such things as float equipment, centralizers, and baskets will cost the production company $11,750. These items will be used in the process of lining the wellbore with protective casing. After the rig is positioned on the pad, additional costs that are covered by the production company include the fuel used to operate the rig and the cost of the various drill bits and reamers used throughout all phases of the drilling operations. The cost of fuel to operate the rig totals on average $32,250, with the cost of the drill bits and reamers totaling $50,000. Further costs include the rental of the instruments and tools that control the direction of the drill bit, which total $45,000. There are also costs for various trucking needs, which total $5,000 and the rental of miscellaneous tools and services for $56,500.

Diesel generators provide all of the power to the drilling sites. These generators are normally provided as part of the leased equipment set with the drilling rig. As many as three 700amp diesel A/C generators power each site. The generators use a variable frequency drive and produce about the same level of power as the power grid provides to a house. Two are typically active at all times, while the third generator is on standby; generators rotate use cycles to prevent overuse and breakdowns.

The fuel used is off-road diesel, a red-dyed tax exempt form of a diesel. It is less expensive than standard diesel, but is of a lower quality. The total diesel expense for a drilling site is approximately $200,000. The diesel expense covers not only the generators but also other diesel vehicles. The generators consume approximately 2,000–3,000 gallons of diesel per day. Fuel costs for generators come to between $50,000 and $75,000 per site. This is based on 2,000–3,000 gallons per day × 25 days (standard drilling period).

Total costs for the drilling of the vertical section of the Marcellus well before drilling even begins average $457,500.

During the drilling of the well, steel tubing, known as casing, is cemented into the ground. During the vertical phase, there are four different sizes of casing that are used. The first section is referred to as the conductor pipe and is generally 20″ in diameter and 20–40 feet long depending on the depth of the first encountered solid rock in the wellbore. The purpose of the conductor pipe is to provide a strong base for construction of the wellbore and the subsequent casing pipe. There is no cement used in the installation of the conductor pipe as it is generally driven into solid rock. The second section of casing is also known as the surface casing and has a diameter of 16¾″. This casing is used to a depth that surpasses the level of the water table. The cost of the surface casing on an average Marcellus well is $19,500. The cement that the casing is surrounded with will cost an additional $15,000. Next the 1st intermediate casing, known as the coal string casing because this casing is used to take the well to a depth past the natural layer of coal that is in the ground, is inserted into the well. The 1st intermediate casing is 11¾″ in diameter and is inserted to a depth of approximately 650 ft and is continued upward until it reaches the surface. The cost of the coal string casing is $12,625, with the cement for this stage adding an additional $10,000. Finally, the 2nd intermediate casing is inserted to a depth of 2,650 ft and once again continued upward until it reaches the surface. The depth to which the 2nd intermediate casing is inserted is much greater than the surface casing and the 1st intermediate casing due to the fact that this is the casing which will reach a point below all possible water aquifers and mines. The cost for this casing runs much higher due to the length of the casing string, $51,500, with the cost of the cement totaling $20,000. After all of the casing has been inserted, a wellhead is placed on well to hold each layer of casing in place. The cost of the wellhead equipment is $5,000.

Throughout the entire vertical drilling operation, the total amount of water used is very minimal in respect to other operations later in the completion process of the well. The only water needed during the vertical drilling phase is used to keep the dust suppressed coming from the wellbore and into the lined cuttings pit during air drilling and also for cementing each casing string. The amount of drilling water needed varies from well to well, but is typically about 500,000 gallons per well. This results in costs for the freshwater for drilling of 500,000 * $ 3 per thousand gallons = $ 1,500, based on a price for the water ranging from $ 3 to $ 15 per thousand gallons, but normally at the lower end. The gas companies pay on a 1,000 gallon basis. Gas companies also have water sharing agreements with other operators to reduce the industry impact.

Depending on the geological characteristics of the location, the water used for drilling may be stored in a pit or in frack tanks. The amount of tanks varies, but is around six on average. The storage tanks are leased. The costs associated with this lease depend on the company and the size of the tank.

Usually, the water used for drilling activities is brought to the location by trucks. The amount of pipeline needed depends on the location of the water source in comparison to the well site. The longest distance they have piped water to a location is approximately five miles (drilling specialist, personal correspondence). The pipelines are rented and charged per foot of pipe rented. As the cost associated with the lease of pipelines is $ 90 per foot, the maximum costs for the pipeline needed in that context is 5 * 5280 feet * $90 = $2,376,000. For the purpose of pumping, a temporary line is used to pump in the water source. As is all the drilling equipment, it is rented.

The costs for the other ingredients of the drilling water, meaning the mud, are approximately $ 7,500–$ 25,000 per well. The amount depends on how much horizontal drilling is necessary. Normally, the drilling mud can be reused for a certain period of time before it begins to break down and needs to be disposed of properly.

Total costs for the drilling of the vertical portion of an average Marcellus Shale in Southwestern Pennsylvania will cost a production company $663,275.

After the drilling of the vertical well has been completed, and the casings have been cemented into place, the vertical well rig is removed from the site.

As with the vertical drilling rig, most production companies do not own their own horizontal drilling rigs and must turn to drilling companies for this stage of the process. The cost of the horizontal drilling rig rental and the labor required to operate the rig average $209,000. Mobilization and setup of the horizontal drilling rig cost $171,000. Once again it is required that a Drill Site Manager be hired to oversee the operation of the horizontal rig, at a cost of $26,500.

Rentals of additional items such as float equipment, centralizers, and baskets will cost the extraction company $15,000. Further costs that are covered by the production company include the fuel used to operate the rig and the cost of the various drill bits and reamers used during the horizontal run. The cost of fuel totals on average $38,000 with the cost of the drill bits and reamers totaling $4,000. There are also costs for various trucking needs, which total $25,000 and the rental of miscellaneous tools and services for $144,750.

On average, the costs incurred by the production company for the setup and operation of the horizontal drilling rig is $633,250.

Horizontal drilling commences at the kick-off point at the bottom of the vertical well. A typical horizontal lateral may be approximately 5,000 feet in length, although in drilling there are variables, such as geology, that effect the drilling decisions. These factors may allow drilling to take the laterals longer, up to as long as 9,000 feet, with a typical decision rule of going "as far as we can laterally while still being economical" (Production specialist, personal correspondence, August 18, 2011).

New technology has enabled drilling rigs to control the drill bit so that they can turn the well from a vertical well into a horizontal well. In order to do this, the extraction company must also rent equipment that is specially designed to control the drill bit as it makes the turn from a vertical direction to a horizontal direction. The cost of this equipment is $85,250. After the drilling is complete, 5½″ casing, known as production casing, is inserted into the well at a cost of $248,500 and secured with $80,000 in cement. Additional costs include $4,000 for the hauling of water used during the cementing process. The wellhead equipment for this stage of the drilling has a total cost of $25,000.

For drilling, it requires some special equipment to separate the drill cuttings from the water. These shakers are included in the rig cost. Disposal of drill cuttings requires about eighty truckloads, which cost about $250 each. One truckload contains 62,000 pound or about 28 metric tons of material (Drilling supervisor, personal correspondence, March 25, 2011). The landfill charges vary per truckload for depositing the cuttings, depending on the landfill used. These charges are impacted by special permissions that landfills need to accept drill cuttings from the Marcellus Shale.

To support both the vertical drilling process and the horizontal well drilling process, there are also costs for drilling mud and chemicals. Drilling mud, which is a combination of water, clay, and various chemicals, is used to float the rock fragments, known as cuttings, and soil back to the surface. This mud is recycled and reused during the course of the drilling of the well. The wellbore is filled with drilling mud just before the vertical rig moves off location to ensure the integrity of the wellbore stays intact and does not collapse while waiting for the horizontal rig to arrive. Costs of filling the vertical portion of the well will cost the production company $10,000, and during the horizontal drilling portion, the cost is $127,800 as much more mud is needed. This mud is recycled after the well is completed and used for the next well drilling operation.

Geologists and engineers play a role at various stages of the drilling. They are not only involved during site selection but also work directly on the drilling rig, collaborating with the drilling crews, to analyze and fine-tune the progress of the drilling. So, in addition to the costs associated with the drilling mud, there are fees paid by the production company to geologists who are employed to complete analysis of the drilling mud and cuttings that are brought to the surface. This process is known as mud logging and enables the crew of the rig to know what geological elements the well is encountering below the surface. This knowledge is important not only to the drilling but to tuning the chemical composition of the drilling mud to best suit conditions at the drilling depth. The cost of this service is $12,000 during the vertical portion, and for the horizontal portion of the well, the cost is $11,050.

Total costs for the drilling of the horizontal portion of a Marcellus Shale well in Southwestern Pennsylvania will cost a production company on average $1,214,850.

At this point, the drilling is complete and the production casing is in place. The horizontal drilling rig is now ready to be deconstructed and moved to another drilling site. These costs are included in the original mobilization costs referenced above.

Drilling an 11,000 foot Marcellus well costs on average $1,878,125. A breakdown of drilling costs is shown in Table 5. In summary, depending on conditions experienced, it takes approximately 18–21 days to drill a Marcellus well.

Various other factors may impact the cost of drilling and fracking, such as the cost of any necessary security measures, if needed. Given the nature of the expensive drilling components, sites may choose to store and secure certain equipment or materials such as drilling bits and expensive parts in secure storage containers, such as CONEX steel storage containers. Each of these containers costs between two to four thousand dollars, and up, depending on size, plus the costs of transportation to the well site. Purchasing security fencing for a well site may cost between $60,000 and $110,000, although fencing rental may cost less.

Phase 4 Hydraulic Fracturing

In the process of hydraulic fracturing, or "fracking," a fracking solution is injected into a well under high pressure. Water, along with additives, fractures the shale rock, while sand props open the fractures, allowing the natural gas to flow (Harper and Kostelnik).

Once the Marcellus Shale well has been drilled and the casing has been inserted and cemented for at least 24 h to cure, it is time to begin the Completions Phase. Completions account for 40–60 % of the overall cost to complete a well. An estimated industry average, per foot, for completions is $500–$600. This amount varies

Table 5 Costs associated with drilling

Vertical drilling	
Surface casing (freshwater): 16–3/4″	$19,500
1st intermediate (coal string): 11–3/4″	$12,625
2nd intermediate casing: 8–5/8″	$51,500
Wellhead equipment	$5,000
Float equipment, centralizers, baskets, etc.	$11,750
Daywork drilling	$225,000
Rig(s) mobilization: all rigs	$32,250
Fuel	$32,250
Bits, reamers, tools, power tongs	$50,000
Pit liners	$24,000
Drilling mud and chemicals	$10,000
Drilling miscellaneous (directional drilling, gyro)	$45,000
Cement surface casing	$15,000
Cement 1st intermediate casing	$10,000
Cement 2nd intermediate casing	$20,000
Trucking	$500
Mud logging	$11,900
Engineering consultant/well-site leader	$25,500
Miscellaneous tools, services, and rentals	$56,500
Haul freshwater for cementing/rig	$5,000
Vertical drilling subtotal	*$663,275*
Horizontal drilling	
Production casing: 5–1/2″	$248,500
Wellhead equipment	$25,000
Float equipment, centralizers, baskets, etc.	$15,000
Daywork drilling: spudder, intermediate, and horizontal rigs	$209,000
Rig(s) mobilization: all rigs	$171,000
Fuel	$38,000
Bits, reamers, tools, power tongs	$4,000
Drilling mud and chemicals	$127,800
Drilling miscellaneous (directional drilling, gyro)	$85,250
Cement production casing	$80,000
Trucking	$25,000
Mud logging	$11,050
Engineering consultant/well-site leader	$26,500
Miscellaneous tools, services, and rentals	$144,750
Haul freshwater for cementing/rig	$4,000
Horizontal drilling subtotal	*$1,214,850*
Total drilling costs	*$1,878,125*

primarily on the length of the lateral and number of engineered stages. If the lateral length is long, there is more length to divide the fixed costs among, thus lowering the price per foot. If the number of stages to be completed is high, there will be additional time and material required to complete the fracturing, thus raising the price per foot. For a 4,500′ lateral Marcellus Shale well, an estimated all inclusive cost can be estimated at $2.5 million, assuming 15 fracturing stages. Hydraulic fracturing companies that provide service to Marcellus Shale play include Halliburton, BJ Services, Baker Hughes, Calfrac Well Services Ltd., and Schlumberger.

The first step in the Completions Phase is to clean out the well. A perforating gun must then be inserted into the well and taken to the very end of the lateral section. These two steps can be done by using a coil tubing rig. On occasion, the perforating gun may be inserted by the directional drilling services, depending on the situation. The cost to initially clean out the well and perforate the first stage can be estimated at $35,000–$50,000. This process, if completed via coil tubing, will require a 3–5 man crew and a coil tubing rig.

Once the first stage has been perforated, the gun is removed and the Fracturing Phase begins. Water is pumped downhole at a rate of 75–100 bpm. This is accomplished with the assistance of 12–18 large water pumps on tractor trailers, circled around the wellhead. All water pumps are connected with highly pressure rated water lines. The water pumps' combined hydraulic horse power is 25,000–30,000. The water is pulled from on-site water completion pits that are capable of holding millions of gallons of water. There are also other means of providing water for fracturing. As water is pumped downhole, casing pressure begins to rise. The pressure required to fracture the Marcellus Shale is between 6,500 and 9,000 psi depending on the formation present. The average is 7,000 psi to stimulate the shale. The water is mixed with additives to create a "fracking fluid," which is pumped downhole, and into the perforations in the casing, made by the perforating gun. The "fracking fluid" squeezes out from perforations in the 4,000- to 8,000-foot-long horizontal arm of the well, which extends through the sedimentary formation, and causes the shale to crack. The shale is tightly compressed and does not release the sought-after quantities of gas until fractured.

Estimated consumption of diesel fuel to complete a single stage by the 12–18 water pumps is 4,000 gallons. Current diesel fuel price for off-road quality is $4 per gallon.

Generally, the amount of fracking water needed varies from well to well. For that reason, different information can be found in this context. Between 4 and 4.5 million gallons and 5.6 million gallons of freshwater per horizontal well are needed for fracking (drilling specialist, personal correspondence; Chesapeake, 2010). Other sources estimate the amount of freshwater necessary for fracking a horizontal well at approximately 3 million gallons (Soeder and Kappel 2009; Airhart 2007). On the contrary, a recent study of Penn State University estimates the freshwater usage for a horizontal well between 4 and 8 million gallons (Abdalla and Drohan 2010). It has been reported that 4 million gallons of water, sand, and chemicals are needed for each well (Hamill 2011). For a vertical Marcellus Shale well, a water consumption of 500,000 to more than 1,000,000 gallons of water is assumed (Harper 2008). Since most of the Marcellus wells are horizontal, for the estimation of the economic impact of a Marcellus well, an assumption of 4 million gallons freshwater usage for the fracking process seems to be reasonable. This would result in costs for the freshwater of 4 million gallons * $3 per thousand gallons = $12,000. Some Marcellus well may need to be hydrofracked several times throughout their productive life (Abdalla and Drohan 2010).

The fracking water is usually stored in one or two pits. The cost for a pit varies depending upon the size of the completion pits, the amount of the overburden that needs to be removed, the terrain, the topography, and other factors. On average, the cost for building a pit is around $120,000 and another $60,000–$70,000 for lining and fencing. There are no real maintenance efforts necessary for the pits other than routine inspections and occasional, minor repairs to the liner.

As for the drilling water, a pipeline is also needed for the transportation of the fracking water. In that context, depending on the distance from the water source to the completions pits that are used to store the water, a few thousand feet to several miles (up to 5 miles) of pipeline are necessary. The pipelines are rented and charged per foot of pipe rented. As the cost associated with the lease of pipelines is $ 90 per foot, the maximum costs for the pipeline needed in that context is 5 * 5280 feet * $90=$2,376,000.

Occasionally, storage tanks are used for the storage of the water in addition to the pits.

The pumps for the frack water are typically rented from water transfer companies. The costs vary depending on the length of the run, how many days the pumps are utilized, and other factors.

Apart from freshwater, the frack fluid includes other ingredients. The costs for those are part of the completion costs that are typically performed by service companies such as Halliburton and BJ Services.

Sand is used during the process to help propagate the fractures and allow gas to flow more easily. Estimated usage of sand is 250 tons per 300 foot stage. The current price of sand, including delivery, is estimated at $4 per ton. This is dependent upon diesel prices and site location. There are various grades of sand that can be used.

Although not widely understood by many, the typical makeup of fracking fluid is available from a number of publicly available sources. Fracking fluid is composed of 92.23 % water and 6.24 % sand, and the remaining 1.54 % makes up the fluid system or additives that aid the efficiency of the fracking fluid (Halliburton 2011). The specific compounds used in any given fracturing operation vary depending on company preference, source water characteristics, and site-specific characteristics, such as the salinity of the deposits. Common components of these include hydrochloric acid (HCl), friction reducers, biocide agents, and scale inhibitors (Halliburton 2011). The total costs for the additional ingredients are between $ 75,000 and $ 200,000.

A small amount (about 10–20 %) of the fracking water flows back, typically within the first 2 weeks after the process, and needs to be disposed of. It is this fracking water that is of environmental concern, as it may contain both fracking solution and brine and other minerals from the well itself. About 10 % of the fracking water flows back during the operation of the well. This water can partly be reused for fracking.

In the context of 220 wells in the Susquehanna River Basin, during the period from June 1, 2008 to May 21, 2010, 59 % of wells used flowback water in fracking, and 88 % of the flowback water brought on-site is used (Abdalla and Drohan 2010). In these 220 wells, the total flowback reused was 44.1 million gallons, while flowback disposed constituted 21.0 million gallons (Abdalla and Drohan 2010).

Besides that, the process for both kinds of water (water from the drilling process and fracking water flows back) is identical. Nevertheless, taking care of the water is a continuous process throughout the entire lifetime of the well, even though the flowback will only be between 5 and 100 barrels per day. As fracking requires 4.5 million gallons of water on average, 450,000–900,000 barrels of water need to be recycled during this period.

The cost for the recycling of both types of water highly depends on the degree of purification desired for the flowback water. The simple disposal of the water costs between $10 and $14 per barrel, although recent regulatory changes have limited water treatment plant's acceptance of Marcellus Shale wastewater. The costs for recycling water range between $3.50 and $5.50 depending on the level of purification achieved. The lower costs refer to water that still contains salt and some minor chemicals and can be reused for the process. The $5.50 version is extremely purified and can be classified as potable.

Several options of achieving recycling or disposal are available. Either a mobile unit that can be placed on-site to limit transportation costs, trucking the water to a wastewater treatment plant, trucking the water to an underground injection site, or building a pipeline system to the plant. The latter option would have the lowest variable costs, but only makes sense if multiple wells exist/are planned in a condensed area. Underground injection is more expensive than recycling, but it is cheaper than treatment (Cookson 2010).

The mobile wastewater treatment unit can either be purchased or rented. Purchasing the equipment (one unit) costs about $4 million, renting $79,500 per month. Additionally, it costs $73,000 to operate it (fuel, labor, etc.). Independent of the option chosen, the costs for water recycling are somewhat similar. The mobile clarifier incurs costs between $2 and $4 per barrel, depending on the level of purification with the lower $2 cost for water that can be reused in the process (Fountain Quail Water Management, www.fountainquail.com, personal communication).

Flowback water requires between 200 and 300 tanker trucks to be shipped for recycling. A well site can choose to recycle this water back into new wells, but this accounts for all flowback being recycled to a separate well site. Recycling saves $200,000 a well and takes 1,000 water trucks off the road (Cookson 2010).

Besides flowback water, other outputs from a wellhead could include garbage and broken materials and equipment. Drilling companies also need to keep the rig clean and measurable, so they work with cleaning companies in the area that have the capabilities to scrub the rig properly in order to allow the engineers to read the measurements on the dials.

The hydraulic fracturing process requires an industry average 25–30-person crew, which includes engineering and maintenance support personnel. Once the first stage has been successfully fractured, a plug is inserted to block water from entering the completed stage and prevent gas from flowing to the surface. Along with the plug, another perforating gun is entered downhole to perforate the second stage.

This can be done via coil tubing or wire line. The plug and gun are lowered to the bottom of the vertical section, but both need to travel to the end of stage one. This can be accomplished by pumping water downhole to carry the plug and gun to the desired location. Once the plug has been set, the perforating gun is discharged. The fracturing process is then repeated. Pumping plugs and perforating guns downhole requires a 3–5-person crew, wire line unit, crane and pressure control equipment. Plugs and perforating guns can be estimated at $5,000–$15,000 each. Labor to perforate one stage and set a plug, on a 400′ stage, is estimated at $15,000–$25,000.

The number of fracturing stages and the length of each stage is engineered specifically to an individual well. Estimated values on a 4,500′ lateral could be 10–20 stages (average 15) and 200′–500′ stage spacing (average 350′). A timeline to complete each stage depends on the operation schedule. For 12 h per day operation, 2–3 stages can be completed. For 24 h per day operation, 4–5 stages can be completed.

The all-inclusive cost per stage to fracture can be estimated at $120,000–$180,000. This price per stage includes all previously mentioned costs (sand, fuel, plugs, perforating gun, services), a portion of the mobilization and demobilization costs ($75,000–$150,000, depending on location) and fracturing services costs (remainder of costs). Additional equipment such as lighting and housing may be required for operations. These items can be rented or purchased by the producing or service companies.

For a Marcellus Shale well with a 4,500′ lateral, the average number of stages can be estimated at 15. The average length of each stage would then be 300′. Using an average of $150,000 per stage to complete ($120,000 + $180,000/2 = $150,000), the total cost to successfully fracture a Marcellus Shale well is $2.5 million.

Phase 5 Completion

Completion of a gas well, over 10–15 days, involves the processes of recapturing flowback and well testing, water recycling (and/or disposal), flare (if needed), and the installation of a "Christmas tree."

Once fracturing is completed, one of the last steps is to drill out the inserted plugs, flow back, and clean out the well. This process can be assumed to cost anywhere from $150,000 to $250,000. For this study, we use the average cost of $200,000, as the actual completion costs at a given well are highly dependent on the site and the amount of reclamation required. Once flowback is complete and enough water has been removed to flow to sales, the well is turned over to production operations to turn the well online.

After the drilling is completed, a piece of equipment with multiple components, consisting of casing head, tubing head, and the "Christmas" tree, is installed at the wellhead in preparation for the controlled extraction of the hydrocarbons from the well. The high pressure of the gases and liquids that are being released from the well requires wellheads that can withstand pressures from 2,000 to 20,0000 psi. Exposure to the weather and potentially corrosive flowback from the well necessitate noncorrosive materials and an ability to withstand temperatures ranging from 50C to 150C. The wellhead must be durable enough to prevent leaking and blowouts caused by high pressure (NaturalGas.org 2010).

Wellhead components and costs are estimated to total between $400,000 and $500,000 (Production engineer, personal correspondence. 24 April 2011). This includes:

- Installation: labor to install all wellhead components costs, approximately $50,000.
- Crushed stone pad: average use of 500 tons at a cost of approximately $30 per ton. Approximate cost is $15,000.
- Casing head: heavy fittings that provide a seal between the well casing and the ground surface. Material is typically steel or steel alloy. Costs can vary from $200,000 to $300,000, depending on the well pressure.
- Tubing head: Tubing head provides a seal between the tubing that is run inside the casing and the ground surface. Its purpose is to provide the connection to control the flow of gas and liquids from the well. Average costs range between $50,000 and $75,000.
- "Christmas tree": the piece of equipment that fits on top of the casing and tubing heads, containing tubes and valves that control the flow of wet and dry hydrocarbons and other fluids out of the well. Its purpose is to allow for the regulation of the production of hydrocarbons from a producing well. A typical Christmas tree is about 4 feet tall (Sweeney et al. 2009) and made of steel or alloy steel. Average cost is $50,000.
- Metering system: Monitors gas production. Average cost ranges between $25,000 and $50,000.

Along with completing the wellhead, land on a well site that is not being used for production but has been disturbed undergoes interim land reclamation. After drilling activity is complete, interim land reclamation is performed based on a plan of operations approved prior to any well development activity commencing. The assessment of site reclamation requirements are based on "the site's habitat quality, quantity of existing habitat, natural features, juxtaposition of those habitats and features on the property, plant and wildlife species currently using the property and those with the potential to use the property based on the habitat present" and can significantly vary (Department of Conservation and Natural Resources 2011).

The approximate site area of a well during development is 300 × 500 feet. During interim reclamation, "40 % of the originally constructed well pad site can be reclaimed. The remaining 60 % of the well pad site is required for maintenance access, produced water storage, and the production equipment noted above." Therefore, the area of the interim land reclamation is approximately 120 × 200 feet (Anderson, Coupal, and White 2009).

Interim reclamation components and costs are estimated to total between $500,000 and $800,000 and are highly dependent on site conditions. These include:

- Recontouring portions of the cleared well site has an estimated cost of $75,000–$150,000, but is very dependent on the topography and amounts of land moved to create the site (Construction Specialist and Production Specialist, personal correspondence, April 12, 2011).

- Reclamation of temporary roads constructed of crushed rock or stone can range between \$180,000 and \$250,000, depending on the site and distance from the main thoroughfares (Construction Specialist and Production Specialist, personal correspondence, April 12, 2011).
- Topsoil spread evenly (estimated 2″ inches) would require approximately 6,912,000 cubic inches (or 40 tons) of soil to reclaim the site. Approximate cost of topsoil is \$20 per cubic yard. Estimated total cost of topsoil would be \$3,000 (CSGNetwork.com 2011). This cost does not include the temporary roadways because distance can be very so significantly. Alternatively, topsoil may be stored during the site construction phase, saving the cost of purchasing topsoil.
- Landscaping and revegetation using a predominately native seed mix to return the land to its natural state. If the area is farm land, it can be seeded only, which usually cost \$30,000–\$50,000. If the area is heavily forested or contains other plant species, the cost of returning the land to its original status can vary widely. There may also be a land owner request where the land owner has requested a certain tree or seed mix when reclaiming (Construction Specialist and Production Specialist, personal correspondence, April 12, 2011).
- Retention pond reclamation, with an average cost of \$15,000–\$25,000 (Construction Specialist and Production Specialist, personal correspondence, April 12, 2011), includes removal of pond liner, backfill, and environmental remediation, which is only required if the pond liner is breached.
- Public road repair is highly dependent on the amount of damage done from well development activities. Some municipalities are planning to request that site operators set aside \$150,000–\$300,000 for public road repairs (Bath 2011) Prototypical road use agreements in Pennsylvania require that on "the completion of the User's operations, the User, at its own cost and expense, shall within 60 days restore the roadways to the same or better condition as existed prior to the commencement of User's operations" (Center for Dirt and Gravel Road Studies 2011). Costs for roadwork can exceed \$500,000 or higher, depending on the site (Construction Specialist, personal correspondence, April 12, 2011).
- Fencing (160 × 300 feet) is typically installed as chain link fence. With each section 6 feet high and 6 feet wide, it rents for approximately \$215 per month, or \$1.42 per 6 foot wide panel. Total rental cost will vary depending on the amount of time the well is producing and the fencing is needed (National Construction Rentals Representative, personal communication, April 26, 2011).

Phase 6 Production

For the purposes of this study, the production stage only covers the gathering system and pipeline. Processing of the natural gas (and potentially other products) is outside the scope of this analysis. There are, however, several requirements within our scope that will be necessary over the 7–15-year lifespan of a well. Costs will include one-time costs such as the finishing off the pad area (typically 300 ft × 500 ft), the gathering pipeline, and interim reclamation costs, such as erosion control, landscape repair, and road repair. Ongoing payments relating to production are royalty payments to the lessor.

Gathering Pipelines

After the drilling and fracturing are done, natural gas begins to flow from the well and pipelines are installed to transport the gas from the wellhead to the market. According to American Petroleum Institute (API), the natural gas pipeline network involves three systems:

- Gathering systems: Production wells are connected through small-diameter pipelines that move pipeline-quality gas from the wellhead directly to the mainline transmission grid. However, since much of the natural gas produced from the Marcellus Shale wells in Southwestern Pennsylvania is "wet gas," it needs to be further refined in a processing plant to remove impurities and natural gas liquids (NGLs) such as propane and butane, before entering the transmission systems.
- Transmission systems "carry the processed natural gas, often over long distances, from the producing region to local distribution systems around the country" (American Petroleum Institute 2011a). The transmission systems consist of 29 % of intrastate pipelines and 71 % of interstate pipelines.
- Local distribution systems: A distribution system, such as local utility, connects to the interstate pipeline at a "city gate" (American Petroleum Institute 2011b). The natural gas is then delivered to homes, businesses, and other end customers.

After securing the rights-of-way, gathering lines are built before production activities begin. Gathering pipelines connect multiple wells, and depending on the production volume, the size of the pipelines ranges between 4 and 24 inches in diameter (Klaber 2010b). Our company interview reveals that the installation and material cost of gathering pipelines is approximately $90 per foot. Therefore, the total cost of gathering pipeline construction is between $95 and $120 per linear foot, with the right-of-way easement being the major contributing factor to the difference.

Table 6 Costs associated with gathering

Gathering pipelines	Likely case
Right-of-way easement	$15
Material and installation	$90
Cost (per foot)	$105
Average length of gathering pipelines for single well (ft.)	4,500
Total cost (per well)	$472,500

Table 7 Natural gas royalty estimates

Natural gas royalty estimate	Likely case
Royalty rate	15 %
Average wellhead gas price	$4.16
Average well production rate (Mcf/per day)	1.3
Acres owned within the well's production unit	640
Number of acres in the well's production unit	640
Expected royalty payment per year	*$296,088*

Source: Adapted from geology.com/royalty/

However, since "the typical Marcellus gathering line has a diameter and a pressure higher than other legacy production and gathering systems within Pennsylvania," some industry data also indicates the economic impact in each mile of new pipeline is approximately $1 million (Klaber 2010a), which is equivalent to $189 per linear foot.

Royalty

The mineral lease agreement between the landowners and the producer is negotiated prior to drilling a gas well. The lease payment, however, "holds the lease on the oil and gas property until drilling and production occur, and thereafter, the lease is held by production until production stops" (Department of Environmental Protection 2010). The minimum royalty on production to the landowners is prescribed by law and set at 1/8, or 12.5 %, of the value of the produced oil or gas (Oil and Gas Leases, 58 P.S. § 33 and § 34). Based on our company interview, royalty ranges from 12.5 % to 25 %, and the current industry average is approximately 15 % (Green 2010).

There are multiple royalty calculators available in estimating the total royalty payment to the landowners (Penn State 2011). Among them, Natural Gas Royalty Estimate (http://geology.com/royalty/) (Geology.com 2011) is widely adopted by the production companies in communicating to the landowners about the projected royalty payment.

The royalty estimate, shown in Table 7, is based on one production unit, which is also the minimum optimal size in drilling a natural gas well. In order to calculate the royalty payment for one production unit, the following assumptions were made: average industry royalty rate of 15 % for Marcellus Shale well, wellhead gas price of $4.16 based on the average of year 2010 (U.S. Energy Information Administration 2011), average well production rate of 1.3 million cubic feet per day (Harper and Kostelnik), and only one well existing in one production unit of 640 acres. The estimate shows that expected royalty payment to the landowners is approximately $300,000 per year. The differences of royalty rate, however, could significantly affect the yearly payment of royalty over the lifetime of well production.

Phase 7 Workovers

Workovers, as part of the ongoing operation of the well, rather than its initial development, are not included in our economic impact analysis. Workover activities could include power generation, such as solar power for the Christmas tree or an on-site generator, additional well stimulation (fracking), equipment maintenance, and servicing.

Phase 8 Plugging and Abandonment/Reclamation

Activities associated with plugging and abandonment of the well and reclamation of the site, such as landscape or road repair, are not addressed within the scope of our economic analysis.

Summary of the Value Chain of a Single Marcellus Shale Well

This study has examined the process of natural gas extraction from the Marcellus Shale, in terms of examining the direct economic impact of a single Marcellus Shale horizontal well site. The spending required in the value chain to bring to production a typical well costs over seven million dollars. These costs are summarized in Table 8.

In summary, while the costs are significant, the development of a Marcellus Shale well is likely to have considerable economic impact on the region. The central costs in development are: site preparation and reclamation (nearly 2/5ths of total cost); mobilization of equipment and materials, including drilling rigs and hydraulic fracking equipment; power generation throughout the process; and steel and steel derivatives. The economic benefits are significant both direct, which this chapter addressed, and indirect and induced economic benefits, not addressed in this chapter.

For some exploration and production firms, they have a reliance on rented or sourced equipment and human resources, allowing individual firms to focus on their

Table 8 Estimated total cost of a Marcellus Shale well

Phase description	
Acquisition and leasing	$2,191,125
Permitting	$10,075
Site preparation	$400,000
Vertical drilling	$663,275
Horizontal drilling	$1,214,850
Fracturing	$2,500,000
Completion	$200,000
Production to gathering	$472,500
Total	*$7,651,825*

core competencies and making available opportunities for specialized entrepreneurial ventures to take part in the value chain.

Government plays a critical role in regulating the industry, and changes to the current laws and regulations are still being considered in Pennsylvania. New regulations or changes to the existing laws and regulations could have a future impact on the costs of drilling and operating a Marcellus Shale well and would certainly impact the value chain as the production companies address issues of compliance. This is one clear example of why the direct economic impact analysis captured in this study is accurate as of the time of the study, but may vary over time in the future as regulatory costs, compliance costs, inflationary pressures, or changes in costs of materials or labor will change the total direct economic impact of a Marcellus Shale well.

Acknowledgments This study was supported by the University of Pittsburgh (Joseph M. Katz Graduate School of Business and the Institute for Entrepreneurial Excellence) and the Washington County Energy Partners, which is "a group of businesses, local politicians and economic development organizations" (Bradwell 2010). We thank the EQT Corporation who made access available to an active drilling site and personnel for data collection.

Michelle K. Bencho, Ian Chappel, Max Dizard, John Hallman, Julia Herkt, Pei Jiuan Jiang, Matt Kerec, Fabian Lampe, Christopher L. Lehner, and Tingyu (Grace) Wei were Katz MBA students who took part in this research project, while Bill Birsic, Emily Coulter, Erik M. Hatter, Donna Jacko, Samuel Mignogna, Nicholas Park, Kaitlin Riley, and Tom Tawoda were undergraduate CBA business students involved in this project. Eric Clements and Roman Harlovic contributed as student intern consultants with the PantherlabWorks Program at the Institute for Entrepreneurial Excellence.

References

Abdalla CW, Drohan JR (2010) Water withdrawals for development of Marcellus Shale gas in Pennsylvania (Marcellus Education Fact Sheet). Penn State College of Agricultural Sciences, Cooperative Extension, University Park. http://extension.psu.edu/water/resources/publications/consumption-and-usage/marcelluswater.pdf

Airhart M (2007) A super-sized thirst. Jackson School of Geosciences, The University of Texas at Austin. http://www.jsg.utexas.edu/news/feats/2007/barnett/thirst.html

American Petroleum Institute (2011a) Natural gas pipelines: introduction. http://www.adventuresinenergy.com/Natural-Gas-Pipelines/index.html

American Petroleum Institute (2011b) Natural gas pipelines: city gates. http://www.adventuresinenergy.com/Natural-Gas-Pipelines/City-Gates.html

Anderson M, Coupal R, White B (2009) Reclamation costs and regulation of oil and gas development with application to Wyoming. Western Economics Forum, Spring 2009. http://ageconsearch.umn.edu/bitstream/92846/2/0801005.pdf

Barth J (2010) Unanswered questions about the economic impact of gas drilling in the Marcellus Shale: Don't Jump to Conclusions. JM Barth & Associates, Inc, Croton on Hudson. http://occainfo.org/documents/Economicpaper.pdf

Bath M (14 Apr 2011) County, drillers nearing accord on road use agreement. Steuben Courier. http://www.steubencourier.com/news_now/x230265204/County-drillers-nearing-accord-on-road-use-agreement

Bradwell M (Aug 2010) Washington county designated as "Energy Capital of the East". Washington County Business Journal. http://www.washingtoncountybusinessjournal.com/stories.php?id=aug10p07s01&month=aug10

Center for Dirt and Gravel Road Studies (2011) Municipal road use agreement. Center for Dirt and Gravel Road Studies, Larson Transportation Institute, Penn State University. http://www.dirt-andgravel.psu.edu/Marcellus/Documents/Sample_RUMA.pdf

Cookson C (2010) Technologies enable frac water reuse. The American Oil & Gas Reporter. Retrieved from http://www.drakewater.com/Oil_Gas_Reporter_March_2010.pdf

Crompton JL (1995) Economic impact analysis of sports facilities and events. J Sport Manage 9(1):14–35

Crompton JL (2006) Economic impact studies: instruments for political shenanigans? J Travel Res 45:67–82

CSGNetwork.com (2011) Top soil cover quantity and cost calculator. http://www.csgnetwork.com/topsoilcalc.html. Accessed 24 Apr 2011

Department of Conservation and Natural Resources (2011) Gas exploration: managing the impacts. http://www.dcnr.state.pa.us/forestry/marcellus/impacts.html. Accessed 24 Apr 2011

Department of Environmental Protection (2010) Landowners and oil and gas leases in Pennsylvania: answers to questions frequently asked by landowners about oil and gas leases and drilling. Department of Environmental Protection (DEP), Commonwealth of Pennsylvania. http://www.elibrary.dep.state.pa.us/dsweb/Get/Document-81979/5500-FS-DEP2834.pdf

Geology.com (2011) Natural gas royalty estimate. http://geology.com/royalty/

Green E (February 28, 2010) Marcellus Shale could be a boon or bane for land owners. Pittsburgh Post-Gazette. http://www.post-gazette.com/pg/10059/1038976-28.stm

Halliburton (2011) Hydraulic fracturing: fluids disclosure. Accessed 10 Aug 2011. http://www.halliburton.com/public/projects/pubsdata/Hydraulic_Fracturing/fluids_disclosure.html

Hamill SD (27 Feb 2011) Scope of Marcellus Shale drilling job creation a matter of conjecture. Pittsburgh Post-Gazette. http://www.post-gazette.com/pg/11058/1128037-85.stm

Harper JA (2008) The Marcellus Shale an old "new" gas reservoir in Pennsylvania. Pennsylvania Geology 38(1):2–13. http://www.dcnr.state.pa.us/topogeo/pub/pageolmag/pdfs/v38n1.pdf

Harper JA, Kostelnik J (n.d.) The Marcellus Shale play in Pennsylvania. Commonwealth of Pennsylvania, Department of Conservation and Natural Resources. http://www.dcnr.state.pa.us/topogeo/oilandgas/Marcellus.pdf

Hefley WE, Seydor SM, Bencho MK, Chappel I, Dizard M, Hallman J, Herkt J, Jiang P J, Kerec M, Lampe F, Lehner CL, Wei T, Birsic B, Coulter E, Hatter EM, Jacko D, Mignogna S, Park N, Riley K, Tawoda T, Clements E, Harlovic R (2011). The economic impact of the value chain of a Marcellus Shale Well. Working paper, Katz Graduate School of Business, University of Pittsburgh. http://d-scholarship.pitt.edu/10484/

Klaber K (2010a) The Marcellus multiplier: expanding Pennsylvania's Supply Chain (October 21, 2010) [PowerPoint slides]. Marcellus Shale Coalition, Canonsburg. http://www.scrantonchamber.com/uploads/press/ShalePresentationPart11287772770.pdf

Klaber K (2010b) Safety and utility oversight of natural gas gathering pipelines in Pennsylvania. [PDF document]. Marcellus Shale Coalition, Canonsburg. http://www.puc.state.pa.us/naturalgas/pdf/MarcellusShale/MS_Comments-MSC.pdf

Louisiana Oil and Gas Association (2008) Horizontal drilling animation. Haynesville Shale Education Center, Louisiana Oil and Gas Association. http://www.loga.la/drilling.html

Miller RE, Blair PD (2009) Input–output analysis: foundations and extensions (2nd revised edition). Cambridge University Press, Cambridge, UK

NaturalGas.org. (2010) Well completion. NaturalGas.org. 2010. http://www.naturalgas.org/naturalgas/well_completion.asp. Accessed 24 Apr 2011

Penn State (2011) Royalty calculator. The Pennsylvania State University, College of Agricultural Sciences Cooperative Extension. http://extension.psu.edu/naturalgas/royalty-calculators

Soeder DJ, Kappel WM (2009) Water resources and natural gas production from the Marcellus Shale (Fact Sheet 20093032). U.S. Geological Survey, U.S. Department of the Interior, Baltimore

Sweeney MB et al (2009) Study guide I, Marcellus Shale natural gas: from the ground to the customer. (The League of Women Voters of Pennsylvania Marcellus Shale Natural Gas Extraction Study 2009–2010). The League of Women Voters of Pennsylvania. http://palwv.org/issues/MarcellusShale/Marcellus%20Shale%20Study%20Guide%20Parts%201-5.pdf

U. S. Department of Commerce (1997) Regional multipliers: a user Handbook for the Regional Input–output Modeling System (RIMS II), Third Edition, March 1997. U.S. Department of Commerce, Economics and Statistics Administration, Bureau of Economic Analysis, Washington, DC. https://www.bea.gov/scb/pdf/regional/perinc/meth/rims2.pdf

U. S. Department of Energy, Office of Fossil Energy, National Energy Technology Laboratory (2009) Modern Shale gas development in the United States: a primer. National Energy Technology Laboratory, Strategic Center for Natural Gas and Oil, Morgantown. http://www.dep.state.pa.us/dep/deputate/minres/oilgas/US_Dept_Energy_Report_Shale_Gas_Primer_2009.pdf

U.S. Energy Information Administration (2011) Natural gas [Data File]. http://tonto.eia.doe.gov/dnav/ng/hist/n9190us3M.htm

Analysis of the Shale Gas Economy in Washington County, Pennsylvania

Yongsheng Wang and Diana Stares

Abstract This study examines the economic potential of unconventional shale gas drilling and production activities in Washington County, Pennsylvania. It fills the research gap of county level economic analysis for Marcellus Shale development in southwestern Pennsylvania, one of the most active areas of shale gas development in America. This study uses the input–output model to analyze drilling and production activities based on public information in order to estimate total economic potential. In addition, the study discusses changes in taxable income from rents, royalties, patents, and copyrights; other taxable income; local housing and rental prices; county real estate tax revenue; sales tax revenue; hotel occupancy; and royalty payments on county-owned lands.

Introduction

Washington County, Pennsylvania, is located in a key area of the Marcellus Shale formation and has experienced significant well development and production in the early years of the Marcellus Shale play. The development of this resource has brought both opportunities and challenges to the local economy. Although Washington County was familiar with extractive industries by virtue of its historical experience with oil and gas development and coal mining activities, the surge of activity presented by the new industry was overwhelming to the county's labor force and economic capacity. As a result, a significant number of workers, engineers, and service professionals from traditional oil- and gas-producing states such as California, Colorado, Louisiana, Oklahoma, and Texas, who had experience working in the unconventional gas industry, were brought to Pennsylvania, along with equipment from those states. The use of

Y. Wang (✉)
Department of Economics and Business at Washington & Jefferson College, 60 S. Lincoln St, Washington, Pennsylvania, USA
e-mail: ywang@washjeff.edu

D. Stares
Center for Energy Policy and Management at Washington & Jefferson College, 60 S. Lincoln St, Washington, Pennsylvania, USA
e-mail: dstares@washjeff.edu

W.E. Hefley, Y. Wang (eds.), *Economics of Unconventional Shale Gas Development*, Natural Resource Management and Policy 45,
DOI 10.1007/978-3-319-11499-6_3

these out-of-state professionals and equipment raises two economic concerns: localization and sustainability. In the context of this study, localization means the proportion of the local spending in Washington County by business entities engaged in the unconventional shale gas industry, their employees, and proprietary income earners (i.e., owners of shale gas companies and owners of shale gas reserves). This is referred to as local purchase percentage (LPP) in regional studies.[1] This development poses two questions: Will a higher level of localization of unconventional gas activities bring more economic benefit? Will a deepened economic relationship with this new industry bring long-term prosperity to the county?

In order to answer the first question and capture the larger picture of this industry, this study presents the economic potential of unconventional gas development in terms of output, tax revenue, and employment. This study is not designed to provide a definitive answer to the second question. However, the findings indicate that in the past decade, local economic activities such as retail sales, hotel rentals, and housing purchases and rentals grew and were positively associated with unconventional shale gas development. The unconventional shale gas development is still in its early stages, and a more pronounced impact may be revealed as the development progresses.

Washington County Economy

Washington County, situated in the southwestern corner of Pennsylvania (PA), is part of the seven county region that forms the Pittsburgh Metropolitan Statistical Area. The county is 857 square miles in size and has a population of 207,820 residents, according to the US Census 2010.

For the first 100 years of its existence, Washington County's economy was largely agricultural based. Its economy experienced dramatic growth from industrial development in the late nineteenth century through the mid-twentieth century, as coal mines, iron and steel mills and related industries were developed and operated in Washington County as part of the steel-manufacturing empire of western PA, as described by the Washington County Planning Commission (2005).

Oil and natural gas also played a part in expanding the Washington County economy beginning in the late nineteenth century. From the 1880s and continuing to the present, oil and gas reserves underlying the county have been developed via conventional wells. The Washington County Planning Commission (2005) notes that in the early 1900s, the county experienced an oil boom with the discovery and production of the McDonald Oil Field in the western part of the county. During this time, PA became the leading producer of oil in the USA. PA continues to produce oil today via conventional wells but at a much reduced rate in comparison with production in other states. In 2011, PA produced approximately 2.7 million barrels of crude oil.[2]

---

[1] Local purchase percentage is defined as the amount of value that is considered to be having an impact on the local economy, according to IMPLAN. http://implan.com/V4/Index.php

[2] Pennsylvania Department of Environmental Protection (2013) Oil and gas well drilling and production in Pennsylvania. http://www.elibrary.dep.state.us/dsweb/Get/Document-94407/8000-FS-DEP2018.pdf. Retrieved 14 July 2014.

The conventional gas industry also continues to the present; from 2001 to 2011, six hundred and fifty-three (653) conventional natural gas wells were drilled in Washington County.[3]

Today, Washington County's economy is largely service based, according to the American Community Survey (US Census Bureau 2011). With a key location at the intersection of two Interstate Highways (I-79 and I-70), 15 miles south of the Greater Pittsburgh Airport, and with a comprehensive transportation system that consists of 2,875 miles of highway (as well as 1,123 miles of state roads and 1,707 miles of local roads), 2 Class 1 railroads, three airports, 40.5 miles of frontage on the Monongahela River, 2 barge lines, 26 terminals, and 7 bus lines, Washington County provides ready access to the cities and markets in Pennsylvania, Ohio, and West Virginia and facilitates the transfer of people and goods by a variety of transportation routes and modes. Thirteen industrial parks promote economic development, and sixteen major shopping centers provide numerous retail sales opportunities. Fourteen public school districts, two institutions of higher education, two community colleges, and three trade/vocational schools educate the county's citizenry to participate in this economy.[4]

Shale Gas Development in Pennsylvania

For the past 10 years, PA has participated in the shale gas development that has revolutionized the country's energy landscape. The use of horizontal drilling and hydraulic fracturing technology has made it economical to access large volumes of oil and gas from shale plays across the nation that were once thought too expensive to exploit, according to the US Department of Energy (2009). The US Energy Information Administration (2011) identifies several shale formations located beneath Pennsylvania, including the Marcellus and Utica Shale formations. The Marcellus Shale has been the focus of development in PA in the past several years. The Marcellus Shale is a Middle Devonian-age black shale situated in the Appalachian Basin. The reservoir spans an area of 95,000 square miles, which extends across most of Pennsylvania and West Virginia and parts of eastern Ohio, southern New York, western Maryland, and western Virginia. PA has the largest share (35 %) of the area of the Marcellus Shale formation.

Following the development of the first successful well in the Marcellus Shale, leasing, permitting, and drilling in the Marcellus Shale accelerated rapidly in PA (Table 1).

[3] Pennsylvania Department of Environmental Protection. Oil and gas reports. http://www.depreportingservices.state.pa.us/ReportServer/aspx?/Oil_Gas/Wells_Drilled_By_County. Retrieved 13 July 2014.

[4] Washington County, http://www.co.washington.pa.us/index.aspx?nid=233. Retrieved 13 July 2014.

Table 1 Number of unconventional wells permitted and drilled in Pennsylvania

	2004	2005	2006	2007	2008	2009	2010	2011	2012	2013
Permitted	6	17	52	181	568	1981	3270	3349	2644	2958
Drilled	2	8	37	115	335	816	1599	1964	1347	1207

Source: PA Department of Environmental Protection, Bureau of Oil and Gas Management, Oil and Gas Reports

The rapid growth of the development is demonstrated by the fact that between 2008 and 2011, PA went from importing 75 % of the natural gas it consumed to becoming a net exporter of natural gas to other states.[5]

Shale Gas Development in Washington County

From the outset, one of the most important areas of the Marcellus Shale play has been Washington County. As with most shale plays, production capabilities are not spread evenly over the Marcellus Shale region. Instead, there are "sweet spots" or areas where wells produce more gas with lower costs, as described by the Governor's Marcellus Shale Advisory Commission (2011). There are two sweet spots in PA – the northeastern counties of Bradford, Tioga, and Susquehanna and the southwestern counties of Washington and Greene. Table 2 shows the tally of unconventional wells permitted and drilled in Washington County in the years from 2004 to 2013. It shows that unconventional wells drilled in Washington County accounted for 43 % of all unconventional wells drilled in PA from 2004 to 2007 and 13 % from 2004 to 2011, while Washington County only accounts for 0.35 % of population and 0.41 % of land area of PA based on the US Census 2010. In 2012, there were 194 unconventional natural gas wells drilled in Washington County. This represents a 26 % increase over the number of unconventional wells drilled in 2011. The number of conventional natural gas wells drilled in 2012 dropped to a single digit (9 wells) for the first time since 2003. Drilling more unconventional wells and fewer conventional wells has been the trend for the prior 5 years in Washington County.

Not only is Washington County a sweet spot and a prime development area for shale gas development, the gas produced in this county differs from that produced in the northeastern sweet spot and further affects the rate of development. The shale gas produced in the northeastern counties is deemed "dry" gas, because it does not

[5] Pennsylvania Department of Environmental Protection (2013) Oil and gas well drilling and production in Pennsylvania. http://www.elibrary.dep.state.us/dsweb/Get/Document-94407/8000-FS-DEP2018.pdf. Retrieved 13 July 2014; Pennsylvania Independent Oil and Gas Association "Traditional Oil and Gas Industry" http://www.pioga.org/publication_file/pioga-traditional-industry-fact-sheet.pdf. Retrieved 14 July 2014.

Table 2 Number of unconventional wells permitted and drilled in Washington County, PA

	2004	2005	2006	2007	2008	2009	2010	2011	2012	2013
Permitted	1	12	26	52	111	210	249	264	325	414
Drilled	0	5	20	45	66	101	166	155	194	220

Source: PA Department of Environmental Protection, Bureau of Oil and Gas Management, Oil and Gas Reports

contain natural gas liquids. The shale gas produced in the southwestern counties is deemed "wet" gas, because it contains natural gas liquids, i.e., ethane, propane, and butane, which must be removed through processing before the gas can be delivered to customers. The natural gas liquids are often more valuable than the methane because of their higher level of British thermal unit (BTU). After they have been separated from the gas, they can be marketed to other users. Ethane is used to make petrochemicals, propane to heat homes and to make petrochemicals, and butane is used in the gasoline refining process.[6] Although dry gas requires less processing before it can be marketed, the extra value added by the liquids make it more profitable to develop the wet gas.[7] According to DEP, in 2012, the unconventional wells in Washington County produced 179,027,481 million cubic feet (Mcf) of natural gas, 1,710,650 barrels of condensates, and 52,239 barrels of crude oil. This was a dramatic increase in production in Washington County from 2011, with a 59 % increase in natural gas and a 223 % increase in condensates. Although the price of natural gas was at a record low in 2012, a large portion of Washington County is located in the wet gas area of the Marcellus Shale formation.[8]

Also distinctive about the economic impacts of the new industry in Washington County is that many of the business entities that are engaged in the unconventional gas industry have located headquarters and branch offices in Washington County, as discussed in Brundage et al. (2011). Of particular note is the development at Southpointe, an industrial park located in Canonsburg, where, in 2011, several upstream companies maintained headquarter/offices as did at least 92 energy-related businesses.[9] By maintaining their offices in Washington County, these companies have significantly enlarged their footprint in the county.

[6] US Energy Information Administration (April 20, 2012) "What are natural gas liquids and how are they used?" http://eia.gov/todayinenergy/detail.cfm?id=5930. Retrieved 6 July 2014.

[7] US Energy Information Administration (8 May 2014) "High value of liquids drives U.S. producers to target wet natural gas resources". http://www.eia.gov/todayinenergy/detail.cfm?id=16191. Retrieved 6 July 2014.

[8] The Map of Wet-Dry Gas, Marcellus Center for Outreach & Research, Penn State University. http://www.marcellus.psu.edu/resources/maps.php. Retrieved on 3 Jan 2014.

[9] Determined by reference to a 2011 map of Southpointe, and the review of the business activities of its residents.

New Economic Opportunities and Challenges

The new industry generated much activity and many changes throughout the PA counties where it conducted well drilling and production activities, creating both opportunities and challenges. One of the most visible signs of the new industry was the large number of laborers working at well sites. As discussed by Brundage et al. (2011), to bring a single Marcellus well on line requires about 420 individuals across 150 different occupations; these individuals collectively perform the labor of 13.1–13.3 full-time equivalent workers. Each subsequent phase of natural gas development, including processing and distribution, has differing workforce needs.

The extraction of the natural gas reserves also brings a new source of income, through lease and royalty payments, to the owners of those reserves. Gamrat (2013) presented the rapid increase of Marcellus Shale royalty payments in PA. Although the amounts of lease and royalty payments are negotiated, the Pennsylvania Oil and Gas Lease Act sets a minimum royalty amount by invalidating a lease which provides for the payment of a royalty less than one-eighth (or 12.5 %) of all natural gas removed or recovered from the royalty owner's property.[10]

In addition to creating private wealth, the development of the Marcellus Shale has the potential to increase public funds. Public entities, such as the state of Pennsylvania, county/municipal governments, state-owned college/universities, and school districts, which own properties that include the gas deposits beneath them, can lease the extraction rights with the same right to royalty payments as private parties.[11] For instance, during the calendar years 2011 and 2012, Washington County received $1.4 million and $1.6 million respectively in royalty payments for natural gas produced from 15 wells drawing shale gas from beneath Cross Creek Park.[12]

Additional public funds arise from the various taxes generated by the unconventional gas development and from the new impact fees created by Act 13, the 2012 statute that governs both conventional and unconventional oil and gas development in PA.[13] The impact fees are paid by the shale gas producers and are distributed, in part, to counties and municipalities hosting shale gas activity for the purpose of mitigating the impacts of Marcellus Shale development on communities.[14] The impact fees for reporting year 2011were distributed in 2012 and totaled $204 million; of this amount, Washington County received approximately $4.4 million (2.08 %), and the county's 66 municipalities collectively received approximately

[10] Pennsylvania Oil and Gas Lease Act, the Act of July 9, 2013, amending the Pennsylvania Guaranteed Minimum Royalty Act of 1979, 58 P.S. Sections 1–5.

[11] Pennsylvania Department of Environmental Protection (2012) Landowners and oil and gas leases in Pennsylvania. http://www.elibrary.dep.state.pa.us/dsweb/Get/Document-91369/8000-FS-DEP2834.pdf. Retrieved 14 July 2014.

[12] Washington County Finance Department.

[13] Pennsylvania Act 13, the Act of February 14, 2012, P.L. 87, No. 13, 58 P.S. Section 2301 et seq.

[14] Chapter 23 of Act 13, 58 P.S. Sections 2301–2316.

$7.2 million.[15] With these payments, Washington County and its municipalities received the third highest amount of impact fees in the state, following Bradford and Tioga counties. The impact fees for the 2012 reporting year were distributed in 2013 and totaled $202 million; of this amount, Washington County received approximately $4.7 million, and its municipalities collectively received approximately $7.9 million.[16] Washington County and its municipalities anticipate receiving annual impact fees for a continuing period of time, with the term and amounts dependent upon the number of new wells drilled in the county and the price of natural gas.

Balanced against these opportunities are a variety of challenges presented by the new industry. These challenges include assuring that PA's work force has sufficient training to avail itself of the employment opportunities offered by this new industry, developing policies that allow for and foster the best use of the shale gas, assuring that the industry addresses the external costs of the development by mitigating impacts on the communities that have supported the shale gas development and enacting sufficiently stringent regulatory standards, particularly with regard to air impacts and surface water and groundwater impacts, to adequately protect the environment and the public health. The ability of PA and Washington County to appropriately respond to these challenges will be critical to the ultimate success of the shale gas industry.

In addition to these challenges, which must be the subject of other studies, is the challenge that Washington County faces on how to assure that the economic benefits created in the county inure to the benefit of the county. The first step Washington County faces in confronting this challenge and assuring its success is to understand the economic potential of the county. That is the purpose of this study.

Data and Methodology

Estimating the Economic Potential of Drilling and Production Activities[17]

This study estimates the economic potential of drilling unconventional natural gas wells and producing gas and related products from unconventional wells in Washington County based on public data. Because it is just the beginning of shale gas

---

[15] Pennsylvania Public Utility Commission http://www.puc.state.pa.us/filing_resources/issues_laws_regulations/act_13_impact_fee_.aspx. Retrieved 16 Jan 2014.

[16] *Ibid.*

[17] In this study, "drilling" refers to the entire process of drilling unconventional wells including site preparation, drilling vertical and horizontal wells, hydraulic fracturing, well completion, and production to gathering. "Production" refers to the process of extracting natural gas and related products from unconventional gas wells and well maintenance. The value is calculated based on wellhead prices minus the cost of depreciation, fee of gathering pipelines, and other amortization charges incurred in previous years.

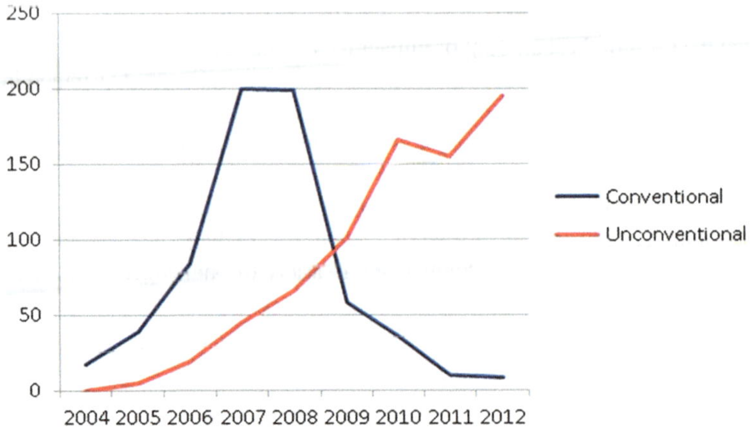

Fig. 1 Total natural gas wells drilled in Washington County (Note: Vertical axis represents the number of wells drilled. "Conventional" indicates conventional wells drilled and "Unconventional" indicates unconventional wells drilled. The horizontal axis represents the calendar year. Data source: Pennsylvania Department of Environmental Protection)

development, it is more important to provide an overview of the county's potential economic capability than to examine what exactly happened in a given year. In order to make these estimates, the project team chose 2011, the first reporting year for the impact fees, as the base of the analysis. Drilling unconventional gas wells and producing gas and related products are the major activities of upstream shale gas companies, and it is the activities of these business entities that are the focus of this analysis. Total economic potential means the combined total of possible direct, indirect, and induced impacts of the drilling and production activities. The direct economic potential is the direct spending on drilling and production activities in the county. The purchases of inputs needed to support the drilling and production activities generate the indirect economic potential. The induced economic potential stems from household spending of wage compensation and royalty payments, i.e., how direct and indirect workers and royalty recipients spend their money.[18]

In order to calculate the direct economic potential of natural gas drilling and production, the project team collected information on total drilling costs and production value from public records such as financial statements and the records of governmental agencies, including the PA Department of Environmental Protection (DEP) and the PA Department of Revenue. Figure 1 shows the annual total number of conventional and unconventional wells drilled from 2004 to 2012 in Washington County.

[18] This study revised the standard induced amount of spending in IMPLAN under different royalty payment rates.

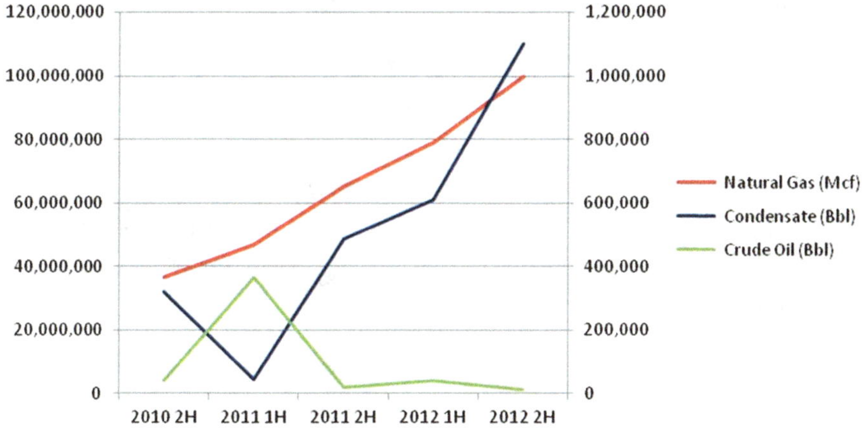

Fig. 2 Production of unconventional wells in Washington County (Note: The vertical axis on the *left* represents natural gas volume measured in thousands of cubic feet (MCF). The vertical axis on the *right* represents condensate and crude oil measured in barrels. The horizontal axis represents half calendar years, e.g., "2010 2H" is the second half of 2010. Data source: Pennsylvania Department of Environmental Protection)

The number of unconventional wells grew substantially during that time period and became the dominant type of new well in Washington County. In 2011, only 10 conventional wells were drilled, whereas 155 unconventional wells were drilled. In terms of production, unconventional wells produce natural gas, condensates, and crude oil. Figure 2 shows the physical production of all three products in Washington County from 2010 to 2012. In 2011, Washington County produced 112,260,435 million cubic feet (Mcf) of natural gas, 529,623 barrels of condensates, and 384,336 barrels of crude oil. With the increase in shale gas drilling and production activities, economic output and employment increased as shown in Figs. 3 and 4.

In order to calculate the indirect and induced economic potential, the project team used IMPLAN (Impact Analysis for Planning), which is a data and software system for economic impact analysis that employs an input–output model.[19] IMPLAN is based on data from the federal Bureau of Economic Analysis and several other federal and state agencies. IMPLAN provides specialized regional economic statistics, which can be used to measure the effect of a given event/industry on a regional or local economy. These potential impacts on the regional economy will be measured in terms of gross output, local government revenues, and employment. In IMPLAN, output is defined as the value of industry production plus any net inventory change. For service sectors, it is the sales value, and it is gross margin for retail and wholesale sectors.

[19] IMPLAN. http://implan.com/V4/Index.php

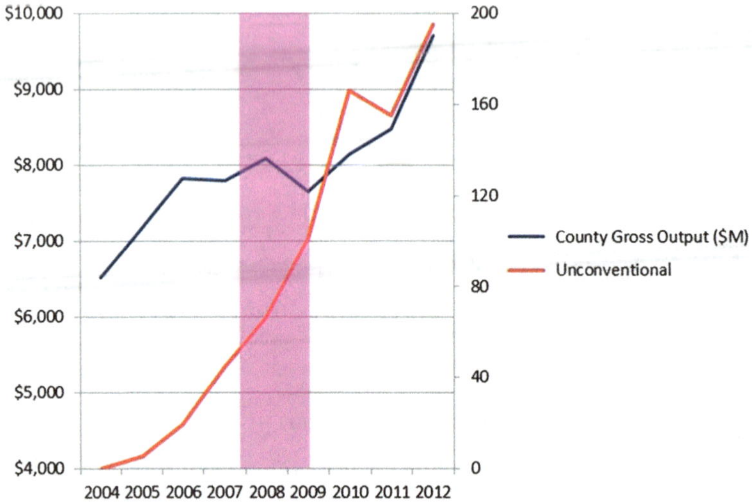

Fig. 3 Gross output of Washington County (Note: The vertical axis on the *left* represents county gross output in millions of dollars. The output number for 2005 is not available. The average between 2004 and 2006 is calculated and used as the proxy. The vertical axis on the *right* represents the number of unconventional wells drilled. The shaded area indicates the most recent national economic crisis. The horizontal axis represents the calendar year. Data source: Pennsylvania Department of Environmental Protection, National Bureau of Economic Research, and IMPLAN)

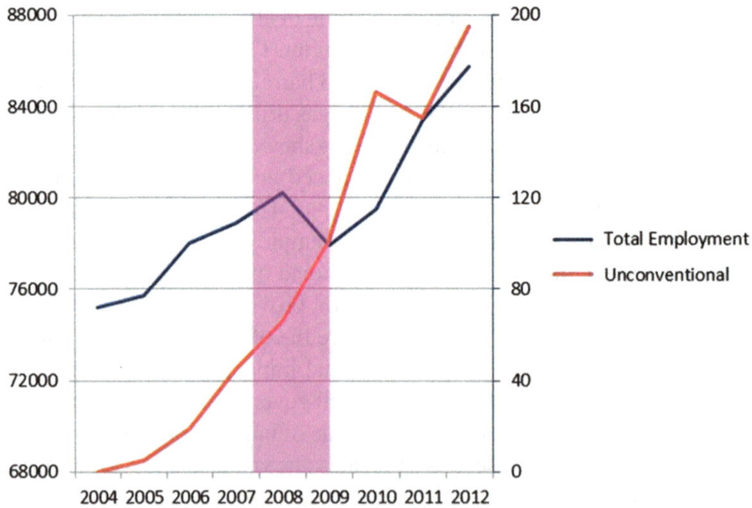

Fig. 4 Employment of Washington County (Note: The vertical axis on the *left* represents total employment in Washington County and the vertical axis on the *right* represents the number of unconventional wells drilled. The shaded area indicates the most recent national economic crisis. The horizontal axis represents the calendar year. Data source: Bureau of Labor Statistics, National Bureau of Economic Research, and Pennsylvania Department of Environmental Protection)

The size of the indirect potential impacts depends on the local availability of inputs, with greater local availability leading to greater indirect potential impacts. The size of the induced potential impacts depends on both the amount of income paid to employees and the amount of royalties paid to owners of natural gas reserves and the amounts both groups spend locally in the county. The more they spend locally (i.e., the less that leaks out of the local economy), the higher the induced potential impacts. In the context of unconventional gas development, royalty payments are a major source of income for natural gas owners that could have a significant potential if spent locally.

To fully understand the potential impacts of economic development on a particular area, it is necessary to understand leakage, or what happens when spending is not localized. Leakage is the amount of income earned in a region and not spent in the region. It includes savings, taxes, and expenditures that occur outside the region. Shale gas development is a new industry in Washington County, and investment and spending have been growing over time. Due to their dynamic nature, it is difficult to estimate the localization rates of spending and investment. In order to provide a full picture and illustrate the future potential of this new industry, this study created two-dimensional tables to illustrate the total economic potential resulting from different combinations of localization rates of royalty spending and drilling activity.

After creating this framework for analysis, the next step was to estimate the economic potential under each combination of localization rates. Typically, a researcher using IMPLAN selects the sector that covers the activity under investigation to analyze the economic potential. However, the default setting in IMPLAN assumes that most of the activities occur locally and that official data (e.g., the data of the Quarterly Census of Employment and Wages from the Bureau of Labor Statistics) for a specific sector accurately reflects these activities. This assumption is generally true when analyzing industries that are traditionally local or analyzing a large geographical area that covers most activities of an industry, e.g., Halaby et al. (2011) and Higginbotham et al. (2010). It is not true for a new industry such as unconventional gas development in a locale like Washington County. Kelsey et al. (2012) reported a similar situation in Bradford County, another drilling county in northeastern PA. Because many drilling activities are subcontracted to drilling companies from other parts of the country, their short-term activities are often reported to governmental agencies in the regions where their company headquarters are located instead of where the work was conducted. In such situations, the traditional single sector method would miscalculate the economic potential. Interviews reflected that many of the outside subcontractors ordered materials, stayed at hotels, and did their personal shopping both locally and outside the county. In order to accurately capture the local potential impacts, this study used a method called analysis by parts with revised LPP in IMPLAN. As indicated by the name, this method analyzes an industry input by input, based on its supply chain, and then combines the potential impact of each individual component with a separate analysis of the spending of employees' wages to calculate the total economic potential of an industry. This approach allows input purchase amounts and locations to be specified by the analyst and thus is often recommended as a way to incorporate the analyst's knowledge of the local economy, as discussed in earlier input–output studies, e.g., Lazarus et al. (2002).

Based on local investigation and Hefley et al. (2011), the study evaluates five major activities of drilling an unconventional gas well: site preparation, drilling, hydraulic fracturing, well completion, and production to gathering. Site preparation involves leveling a site, implementing erosion control, mobilizing equipment, and constructing access roads. Drilling, well completion, and hydraulic fracturing require five common components: water, fuel, labor, specialized materials, and specialized equipment. Water is obtained locally or recycled; the main cost of obtaining the water is transportation. Equipment on the drilling site uses tax-exempt red dyed diesel as mentioned in Hefley et al. (2011). Both fuel and specialized materials can be purchased either locally or shipped from outside. Specialized equipment is either purchased or rented. Production to gathering is the final step before production.[20] In addition to purchasing materials and installation, companies need to pay right-of-way easements to the owner of surface lands whose properties must be accessed to conduct these activities. The economic potential from drilling depends mainly on how much of the above activities are localized in the county.

Production consists of two major activities: extraction and well maintenance. Most of the work associated with production activities is performed by local employees. Also, as discussed above, many gas-producing companies in southwestern PA have established either headquarters or branch offices in Washington County. These offices manage the operations of these companies in Washington County, as well as in nearby counties in the region, and in other areas related to gas development in the eastern part of the USA.

A final element of the production costs is the royalty payment. There is no direct information available on county level royalty income because royalty payments are privately negotiated. There are many factors involved in determining royalty rates, and they vary from property to property. The key factors are the size of a property, ease of access, environmental constraints, as well as potential production rates. As discussed above, the Pennsylvania Oil and Gas Lease Act, 58 P.S. Section 33 et seq., requires natural gas producers to pay natural gas owners a minimum of 12.5 % royalty based on the quantity of gas produced and the market price. The PA State Supreme Court ruled in 2010 that gas companies could deduct postproduction costs before paying royalties.[21] Thus, the gas price with pipeline fees deducted is a reasonable value for calculating royalties.

Royalty income amounts can be indirectly assessed by reviewing changes in royalty income tax payments as shown in Fig. 5. PA tax returns group together the income from rents, royalties, patents, and copyrights (RRPC), which indirectly reflects changes in royalty income. According to the PA Department of Revenue,

[20] The cost of production to gathering in the context of this study includes the payment of right-of-way easements, material cost and installation, and cost of short-distance pipelines before a processing plant or main transmission grid.

[21] Refer to Kilmer v. Elexco Land Services, Inc., 990 A. 2d 1147 (PA 2010).

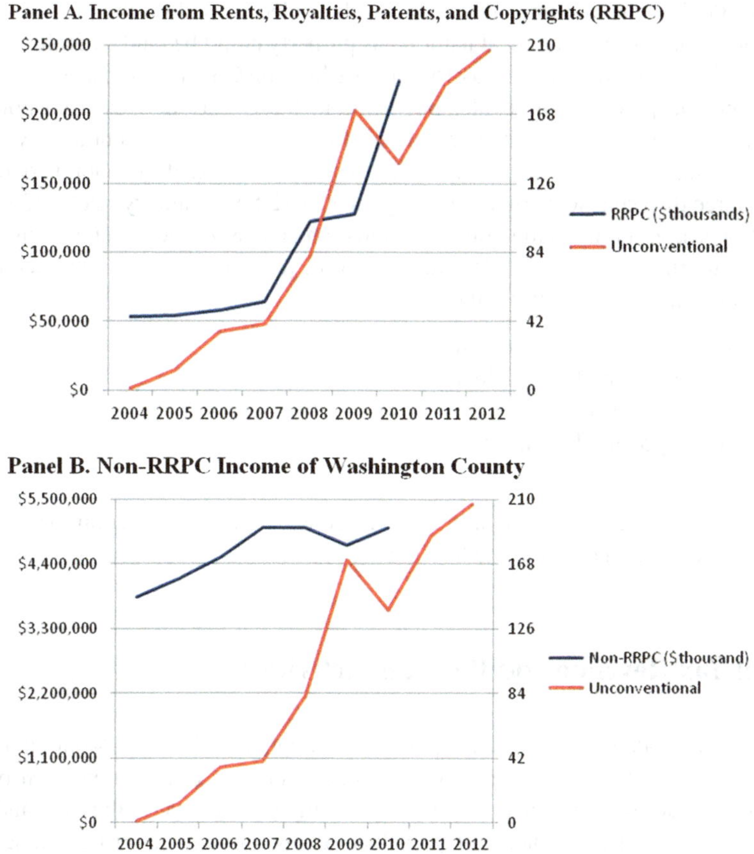

Fig. 5 Taxable income of Washington County. *Panel A. Income from rents, royalties, patents, and copyrights (RRPC)* (Note: The vertical axis on the *left* represents RRPC, total taxable income of rents, royalties, patents, and copyrights from Washington County measured in thousands of dollars. The vertical axis on the *right* represents the number of unconventional wells drilled. The horizontal axis represents the fiscal year. The Pennsylvania state fiscal year runs from July to June. All numbers in the graph are adjusted accordingly. Data source: Pennsylvania Department of Revenue and Department of Environmental Protection). *Panel B. Non-RRPC income of Washington County* (Note: The vertical axis on the *left* represents non-RRPC, total taxable income from Washington County excluding the income of rents, royalties, patents, and copyrights measured in thousands of dollars. The vertical axis on the *right* represents the number of unconventional wells drilled. The horizontal axis represents the fiscal year. The Pennsylvania state fiscal year runs from July to June. All numbers in the graph are adjusted accordingly. Data source: Pennsylvania Department of Revenue and Department of Environmental Protection)

there was a 322 % increase in RRPC in Washington County between 2004 and 2010 as compared to the PA state average of 72 % during the same period. The category of income that most likely caused this large difference between the county and state averages was royalty income. The correlation coefficient between RRPC income and the amount of unconventional wells drilled is 0.81, while it is 0.63 between non-RRPC income and unconventional wells drilled. Based on the project team's interviews, an owner of an ideal natural gas property could be paid a royalty as high as 18 %, but the common rate was 15 % in Washington County. As discussed earlier, the economic potential of royalty payments in the county derives from both the amount of royalty paid and the percentage of the royalty that is spent locally.

Thus, rather than use one royalty rate, this study evaluates the economic potential under three differing scenarios of royalty rates: a 12.5 % statutory rate, a 15 % rate found to be commonly paid in our interviews, and an 18 % rate which is considered a high rate in the county. Under each scenario, this study presents the economic potential from six perspectives including:

- Total economic potential on output
- Economic potential vs. the local investment
- State and local tax revenue
- Economic potential vs. tax revenue
- Employment

As noted above, a 15 % royalty rate was considered the most common rate and will be discussed in detail in the findings.

Local Tax Revenue and Business Activities

In addition to calculating economic potential through use of the input–output model, this study collected a large amount of business and tax information from public sources, including hotel occupancy rates, housing market data (both rental and purchases), and local retail data to reflect local economic characteristics along with shale gas development. Some of the information came from the Washington County government, which follows a calendar fiscal year, and some of it came from PA state government, which follows a July-to-June fiscal year; the drilling information that was used was adjusted accordingly.

The study reflected that when out-of-state contractors moved into the county to work, their employees stayed in local hotels, rented apartments, and even purchased single-family homes – depending on how long they planned to stay in the county. As shown in Fig. 6, the correlation coefficient between hotel occupancy tax revenue and unconventional wells drilled is 0.97. According to the Washington County Finance Department, there was an approximately 200 % increase in hotel occupancy tax revenue between 2004 and 2012. Washington County charged a fixed 3 % tax on hotel services over the research period. The change in tax revenue reflected directly the change in hotel revenue.

Between 2004 and 2012, there was increasing demand in the rental housing market in Washington County. Figure 7 shows rental prices for apartments and condos in Panel A and for single-family homes in Panel B. Panel A indicates that the rental of three- and four-bedroom apartments and condos is more associated with the development of unconventional wells than is the rental of apartments and condos

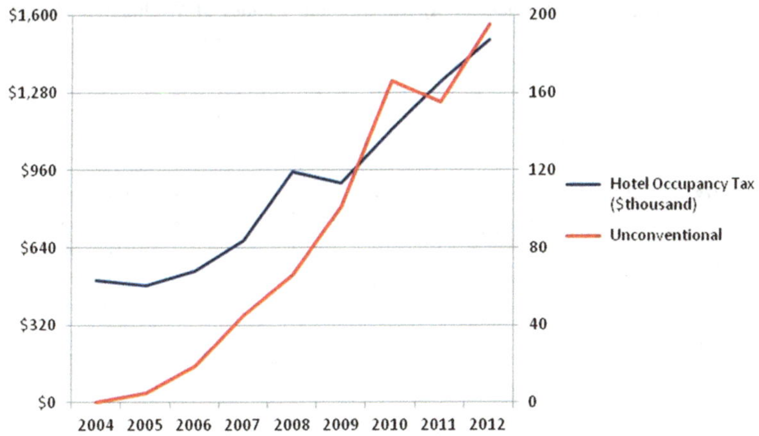

Fig. 6 Hotel occupancy tax revenue collected by Washington County (Note: The vertical axis on the *left* represents hotel occupancy tax collected by Washington County measured in thousands of dollars. The vertical axis on the *right* represents the number of unconventional wells drilled. The horizontal axis represents the calendar year. Data source: Washington County Finance Department and Pennsylvania Department of Environmental Protection)

with fewer bedrooms. This may be due to the fact that these types of dwellings are more convenient for families to stay in or workers to share. The rental market for single-family homes follows a similar pattern. Three- and four-bedroom houses are more associated with well development than are houses with fewer bedrooms.

Figure 8 shows housing prices in Panel A and county real estate tax revenue in Panel B. Housing prices grew in step with the increase in unconventional wells drilled before the latest national economic crisis beginning in late 2007. As expected, prices went down during the crisis. Surprisingly, the local housing market recovered several months before the official ending of the crisis, the most severe economic crisis in the USA since the Great Depression. However, shale gas development before 2007 was very limited in the county. The graph cannot provide the definitive description of this relationship. Gopalakrishnan and Klaiber (2012) and Muehlenbachs et al. (2012) give a more detailed analysis of housing prices and unconventional well development in Washington County. Lipscomb et al. (2012) provide a general understanding of housing values and unconventional well development. County real estate tax revenues increased steadily also. Although there was a rate increase in 2009, for comparison purposes, this study kept the rate constant over the research period.

The final area of interest is retail sales. There are three categories of retail sales based on sales tax collection: motor vehicle (MV) sales, non-motor vehicle (NMV) sales, and wine and liquor (WL) sales. Any local spending from an industry boosts the local economy even if it is only partially spent in local establishments. Figure 9 shows sales tax revenue collected from Washington County by the PA state government including both MV sales tax and NMV sales tax. Washington County had a fixed 6 % sales tax over the research period. The change in sales tax revenue directly

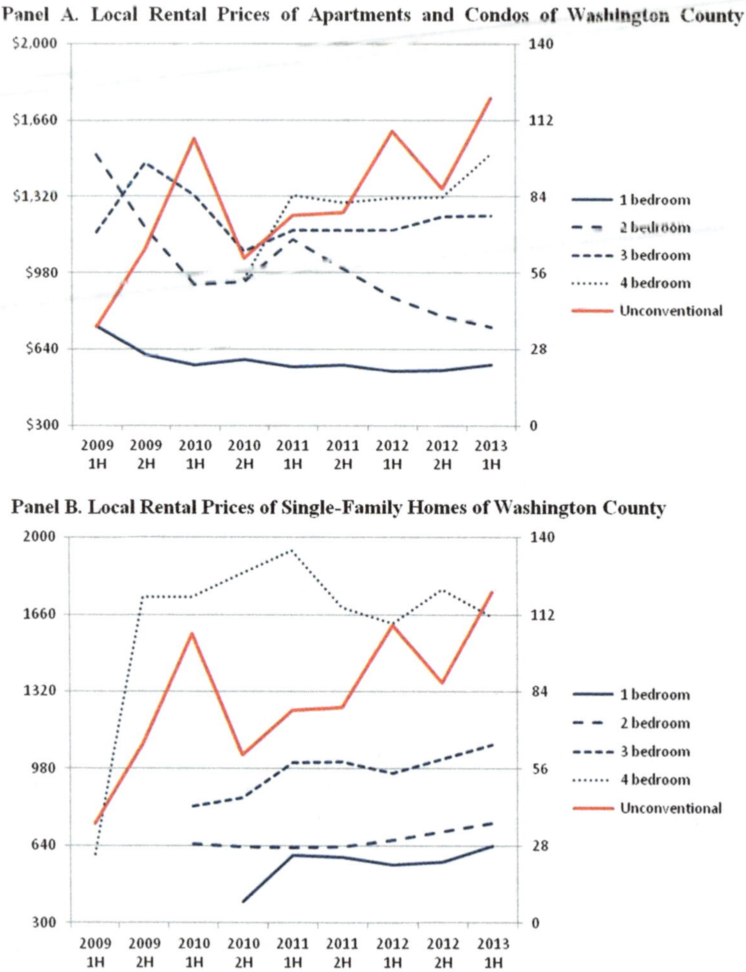

Fig. 7 Local housing rental market. *Panel A. Local rental prices of apartments and condos of Washington County* (Note: The vertical axis on the *left* represents median rental prices of apartments and condos with one to four bedrooms. The vertical axis on the *right* represents the number of unconventional wells drilled. The horizontal axis represents the half calendar year, e.g., "2009 1H" is the first half of 2009. Data source: RentRange and Pennsylvania Department of Environmental Protection). *Panel B. Local Rental Prices of Single-Family Homes of Washington County* (Note: The vertical axis on the *left* represents median rental prices of single family homes with one to four bedrooms. The vertical axis on the *right* represents the number of unconventional wells drilled. The horizontal axis represents the half calendar year, e.g., "2009 1H" is the first half of 2009. Data source: RentRange and Pennsylvania Department of Environmental Protection)

reflects the change in sales. Sales tax in both the MV and NMV categories had sharp increases in the middle of the 2009 state fiscal year, which coincided with the end of the economic crisis and the increased pace in well drilling. There was a more than 100 % increase in the number of unconventional wells drilled during the 2009 fiscal

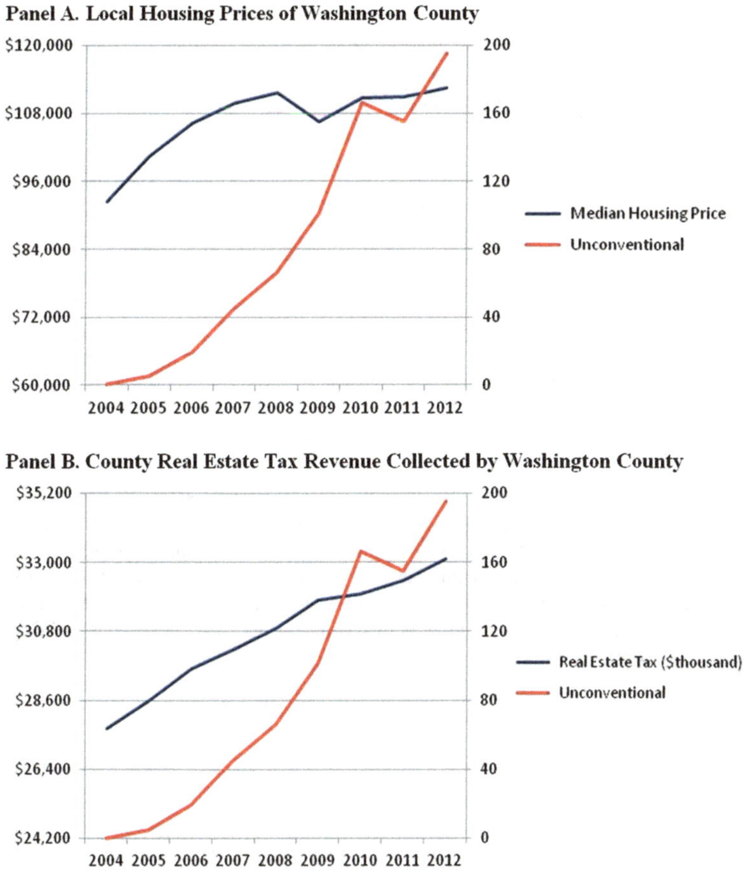

Fig. 8 Local housing market. *Panel A. Local housing prices of Washington County* (Note: The vertical axis on the *left* represents median price of single family home in Washington County. The vertical axis on the *right* represents the number of unconventional wells drilled. The horizontal axis represents the calendar year. Data source: DataQuick and Pennsylvania Department of Environmental Protection). *Panel B. County real estate tax revenue collected by Washington County* (Note: The vertical axis on the *left* represents real estate tax collected by Washington County measured in thousands of dollars. To maintain consistency, this study excluded the millage increase after 2009. The millage is kept at 21.4 throughout the sample period. The vertical axis on the *right* represents the number of unconventional wells drilled. The horizontal axis represents the calendar year (Data source: Washington County Finance Department and Pennsylvania Department of Environmental Protection)

year compared to the prior year. 2009 was the first fiscal year in which the level of drilling in Washington County reached 170 unconventional wells, and in the subsequent three fiscal years, there were annual averages of more than 170.

In PA, the state government regulates sales of wine and liquor (WL). NMV sales tax revenue does not include tax collected on WL sales but does include tax collected on sales of beer. Figure 10 shows WL sales in Washington County experienced a steady

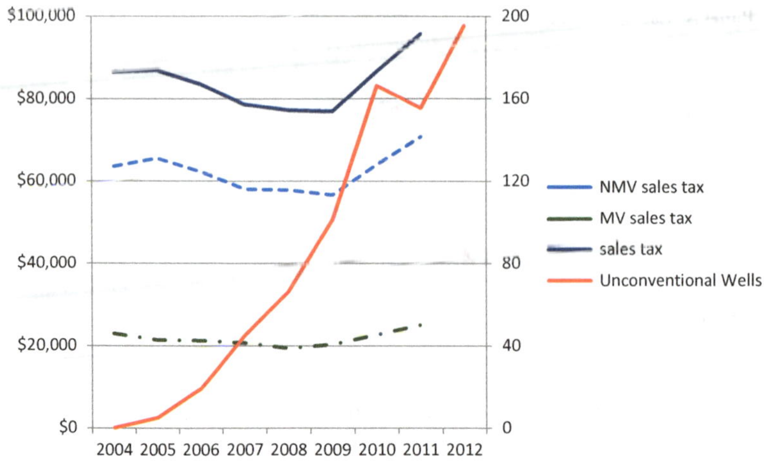

Fig. 9 Sales tax revenue from Washington County collected by PA (Note: The vertical axis on the *left* represents non-motor vehicle (NMV) sales tax, motor vehicle (MV) tax, and total sales tax collected by PA state government from Washington County measured in thousands of dollars. The vertical axis on the *right* represents the number of unconventional wells drilled. The horizontal axis represents fiscal year. The Pennsylvania state fiscal year runs from July to June. All numbers in the graph are adjusted accordingly Data source: Pennsylvania Department of Revenue and Department of Environmental Protection)

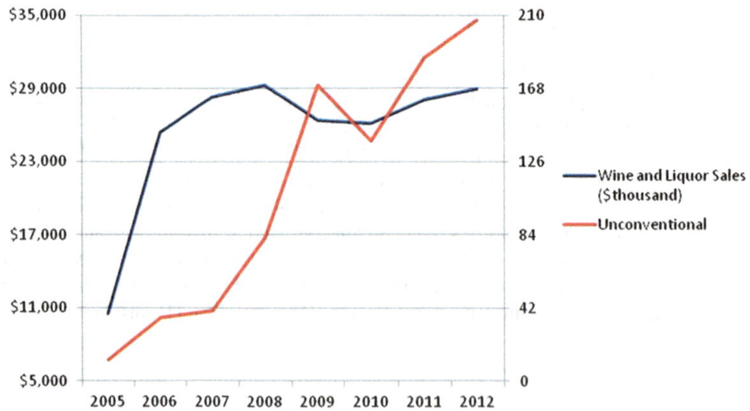

Fig. 10 Wine and liquor sales from Washington County collected by PA (Note: The vertical axis on the *left* represents wine and liquor sales in Washington County measured in thousands of dollars. The vertical axis on the *right* represents the number of unconventional wells drilled. The horizontal axis represents fiscal year. The Pennsylvania state fiscal year runs from July to June. All numbers in the graph are adjusted accordingly Data source: Pennsylvania Liquor Control Board and Department of Environmental Protection)

increase prior to the economic crisis and suffered during the crisis. It was still below precrisis levels during fiscal year 2011. Thus, WL sales had a much lower level of correlation with well drilling than other categories of retail sales in the county.

Research Findings

This section presents the findings of the input–output model under the scenario of 12.5 %, 15 %, and 18 % royalty payment rates.[22] For the simplicity of discussing the findings, the results in the case of a 50 % local royalty spending rate and 50 % drilling activity localization rate (50/50 localization rate) will serve as the main example, followed by the variations of 25/25 localization rate and 100/100 localization rate. Table 3 shows the total economic potential under the 50/50 scenario.

It shows that with an average of a 15 % royalty payment, drilling and production activities of unconventional gas producers in Washington County had a total potential impact of 9.01 % on the local economy in 2011 – assuming the 50/50 localization rate. Under the case of a 25/25 localization rate, the potential would be 6.64 %. Under the 100/100 rate, the potential would be 13.75 %. The higher the localization rates, the greater the impact on the local economy. There is no information available on the exact localization rate in Washington County. Understanding the variation of economic potential under different localization rates will help the industry and interested stakeholders to better understand the county's economic potential and to design policies for local development.

Table 3 Total economic potential of drilling and production on output in Washington County under the 50/50 localization rate as a percentage of total output

Royalty rate / Year	12.5 %	15 %	18 %
2011	8.97 %	9.01 %	9.06 %

With a 15 % royalty rate, the per dollar potential of local spending from unconventional gas development in Washington County as shown in Table 4 is 142 % of the local spending under a 50/50 localization rate. In cases where the localization rate ranges from 25/25 to 100/100, local investment could generate an impact between 139 % and 146 % of the local spending.

[22] Numerical findings presented in this section are based on the 2011 price.

Table 4 Total economic potential from drilling and production on output of Washington County under the 50/50 localization rate as a percentage of total local spending

Year \ Royalty rate	12.5 %	15 %	18 %
2011	141 %	142 %	143 %

Tables 5 and 6 show that out of the total economic potential, there could be 6.27 % contributing to state and local tax revenues in the case of a 50/50 localization rate, and these newly generated tax revenues could account for 9 % of total local spending in drilling and production.

Table 5 Percentage of the economic potential contributing to state and local tax revenue under the 50/50 localization rate

Year \ Royalty rate	12.5 %	15 %	18 %
2011	6.24 %	6.27 %	6.31 %

Table 6 Economic potential contributing to state and local taxes under the 50/50 localization rate as a percentage of total local spending in drilling and production

Year \ Royalty rate	12.5%	15 %	18 %
2011	8.83 %	8.91 %	9.02 %

Table 7 reveals the employment potential. In the case of a 50/50 localization rate, 5.01 % jobs at the county could be directly and indirectly created through unconventional gas-related activities. Also, the higher the localization rates, the lower the cost to create jobs.

Table 7 Employment potential of drilling and production under the 50/50 localization rate as a percentage of total employment in Washington County

Year \ Royalty rate	12.5 %	15 %	18 %
2011	4.97 %	5.01 %	5.06 %

When estimating the economic potential of a new industry entering a local economy, there are at least two possible concerns regarding the magnitude of the potential: displacement and leakage, as discussed in Weinstein and Partridge (2011). The displacement effect means that the new industry could potentially attract workers from the existing industries without contributing to overall local employment growth.

Figure 4 shows that local employment growth positively associates with the number of unconventional wells drilled before the 2007 economic crisis. Following the crisis, the average annual employment growth rate between 2009 and 2012 doubled as compared to the rate between 2004 and 2007, which coincided with a surge in the number of unconventional wells drilled. Although there is a positive impact in terms of total employment, there is no definitive answer regarding a displacement effect due to data unavailability. Closely monitoring the local economy in the future is necessary to understand the sustainability of the economic potential.

Regarding leakage, it is hard to measure or predict the dynamic spending patterns of the employees and of royalty earners – their primary spending could be in Washington County or elsewhere. For example, under a 15 % royalty payment rate and 50 % drilling localization rate, if the local spending rate of royalty income is increased from 25 % to 100 %:

- The total economic potential could range between 8.88 % and 9.27 % of total output.
- The total economic potential could range between 140 % and 146 % of total local spending in drilling and production.
- The percentage of total economic potential contributing to state and local tax revenues could range between 6.18 % and 6.44 %.
- The total economic potential contributing to state and local tax revenues could range between 8.66 % and 9.42 % of total local spending.
- The employment potential could range between 4.89 % and 5.25 % of total employment in Washington County.

Although the pattern is unpredictable, the above findings provide certain policy implications from an economics perspective. If Washington County wants to retain the benefit of this industry, it must provide opportunities and an environment that encourages individuals to spend money locally and unconventional gas producers to localize their operations. Of course, there are issues beyond economics for policy makers to consider in making such important decisions.

Conclusion

Its geographical location and long history of oil, gas, and coal development give Washington County, PA, a unique position in the development of unconventional shale gas. Both opportunities and challenges face this new development, from economic, geopolitical, social, environmental, and public health standpoints. One of the key challenges is to assure that the benefits of this industry are sustainable. This study has focused on the economic potential of drilling and production activities of unconventional gas producers in Washington County. The key to sustainable economic development for generations to come is to retain locally the benefits generated by this industry. Although we do not know the current level of localization, the collected data reflects early signs of positive impacts. By increasing both the local spending rate of royalties and the localization rate of drilling activities from 25 % to 100 %, the economic potential of drilling and production in Washington County could increase from 6.64 % to 13.75 % of the total output, assuming an average royalty payment rate of 15 %. The potential impact on jobs could increase from 3.41 % to 8.20 % of total employment. Now, the challenge is to increase spending and development activities locally. This study reflects the beginning of our understanding of this new industry in Washington County. From an economic standpoint, one of the future research tasks suggested by this study is to measure the localization rate of unconventional shale gas development by conducting surveys among royalty earners and industry suppliers. To have a full picture of all aspects of the development, we need to take a holistic approach, which can be best accomplished through cross-discipline collaboration.

Acknowledgments This study was supported by Center for Energy Policy and Management at Washington and Jefferson College, Washington County Energy Partners, and MSC Foundation.

References

Brundage TL, Jacquet J, Kelsey TW, Ladlee JR, Lobdell J, Lorson JF, Michael LL, Murphy T (2011) Pennsylvania statewide Marcellus Shale workforce needs assessment. Marcellus Shale Education and Training Center, Penn State University

Gamrat F (2013) Marcellus royalty payments rising rapidly. Allegheny Institute for Public Policy, Policy Brief 13(27). http://www.alleghenyinstitute.org/wp-content/uploads/components/com_policy/uploads/Vol13No27.pdf

Gopalakrishnan S, Klaiber HA (2012) Is the Shale boom a bust for nearby residents? Evidence from housing values in Pennsylvania. Am J Agric Econ 96(1):43–66

Governor's Marcellus Shale Advisory Commission (2011) "7/22/2011 Report" Office of the Lieutenant Governor, Commonwealth of Pennsylvania. http://files.dep.state.pa.us/PublicParticipation/MarcellusShaleAdvisoryCommission/MarcellusShaleAdvisoryPortalFiles/MSAC_Final_Report.pdf

Halaby D, Oyakawa J, Shayne C, Keairns C, Dutton A (2011) Economic impact of the eagle ford Shale. Institute for Economic Development, Center for Community and Business Research, University of Texas at San Antonio

Hefley WE, Seydor SM, Bencho MK, Chappel I, Dizard M, Hallman J, Herkt J, Jiang P, Kerec M, Lampe F, Lehner CL, Wei T, Birsic B, Coulter E, Hatter EM, Jacko D, Mignogna S, Park N, Riley K, Tawoda T, Clements E, Harlovic R (2011) The economic impact of the value chain of a Marcellus Shale Well. Joseph M Katz Graduate School of Business, University of Pittsburgh, Working paper

Higginbotham A, Pellillo A, Gurley-Calvez T, Witt TS (2010) The economic impact of the natural gas industry and the Marcellus Shale Development in West Virginia in 2009. Bureau of Business and Economic Research, College of Business and Economics, West Virginia University

Kelsey TW, Shields M, Ladlee JR, Ward M (2012) Economic impacts of Marcellus Shale in Bradford County: employment and income in 2010. Marcellus Center for Outreach & Research, Penn State University

Lazarus WF, Platas DE, Morse GW, Guess-Murphy S (2002) Evaluating the economic impacts of an evolving Swine Industry: the importance of regional size and structure. Rev Agric Econ 24(2):458–473

Lipscomb CA, Kilpatrick SJ, Wang Y (2012) Unconventional Shale gas development and real estate valuation. Rev Reg Stud 42(2):161–175

Muehlenbachs L, Spiller E, Timmins C (2012) Shale gas development and property values: differences across drinking water sources. NBER Working paper

U.S. Census Bureau (2010) Census of population. U.S. Department of Commerce, Washington, DC

U.S. Census Bureau (2011) 2007–2011 American Community Survey 5-Year Estimates. U.S. Department of Commerce, Washington, DC

U.S. Department of Energy (2009) Modern Shale Gas in the United States: a primer. http://www.netl.doe.gov/technologies/oil-gas/publications/EPReports/Shale_Gas_Primer_2009.pdf

U.S. Energy Information Administration (2011) Review of emerging resources: U.S. Shale Gas and Shale Oil Plays. http://eia.gov/analysis/studies/usshalegas/pdf/usshaleplays.pdf

Washington County Planning Commission (2005) Washington County Comprehensive Plan, Adopted November 23, 2005. http://www.co.washington.pa.us/downloads/Washington_County_Comprehensive_Plan.pdf

Weinstein AL, Partridge MD (2011) The economic value of Shale natural gas in Ohio. Ohio State University, Working paper

The Shale Gas Economy in the Northeast Pennsylvania Counties

Kirsten Hardy and Timothy W. Kelsey

Abstract More than half of the Marcellus Shale wells drilled in Pennsylvania through 2013 are within six counties of the Northern Tier, which account for four of Pennsylvania's top five drilling counties. These counties overwhelmingly are rural, with low population densities and relatively small economies. Unlike the Marcellus Shale counties in southwest Pennsylvania, they lack past gas and coal development, so the activity is fundamentally different than what they have experienced. This chapter explores the economic impacts occurring within the Northern Tier counties. Due to their relatively small population size, such changes are relatively more visible than if the economies were larger.

Introduction

Of the 7,432 unconventional wells drilled by the end of 2013 in Pennsylvania, over half were located within the six counties of the Northern Tier (DEP 2014), which include Bradford, Lycoming, Sullivan, Susquehanna, Tioga, and Wyoming counties. Drilling activity increased rapidly within the first 7 years of shale development (2005 through 2011), during which time 2,694 unconventional gas wells had been drilled in this region (DEP 2014) (Fig. 1 and Table 1) and the activity has had profound impacts on the region's local businesses, residents, workforce, and property owners. The rate of drilling fell slightly in 2012 and 2013 due to falling gas prices and a new focus on drilling in Ohio. Still, in those 2 years over 2,500 wells were drilled across the state and over 1,300 wells were drilled in the Northern Tier.

Despite the level of drilling activity, it is unclear how much of the economic impact from Marcellus Shale development is actually occurring in the counties with

K. Hardy (✉)
Former student in Community, Environment and Development,
The Pennsylvania State University, University Park, USA
e-mail: kirstenhardy81@gmail.com

T.W. Kelsey
Department of Agricultural Economics, Sociology and Education,
The Pennsylvania State University, University Park, USA
e-mail: tkelsey@psu.edu

W.E. Hefley, Y. Wang (eds.), *Economics of Unconventional Shale Gas Development*, Natural Resource Management and Policy 45,
DOI 10.1007/978-3-319-11499-6_4

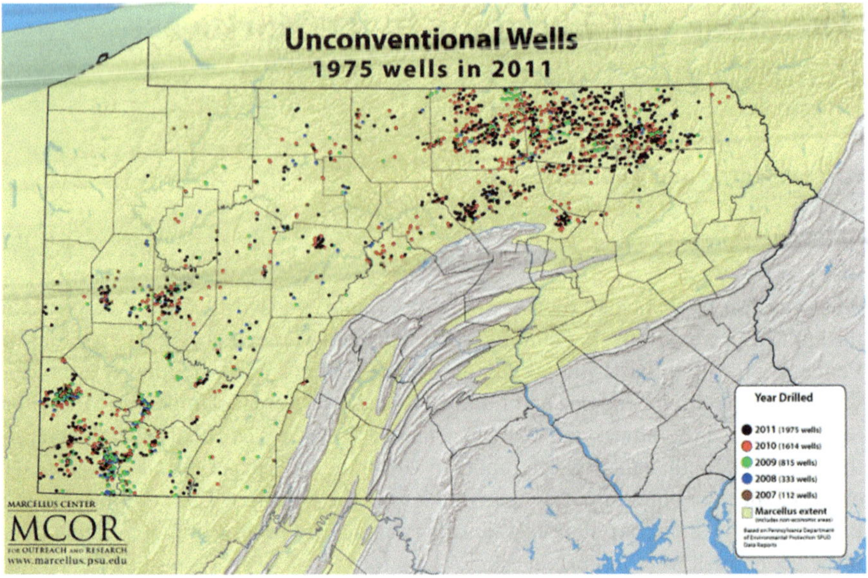

Fig. 1 Marcellus Wells in Pennsylvania, 2007–2011. (Source: Penn State Marcellus Center for Outreach and Research)

Table 1 Marcellus Wells drilled, by year

	2004	2005	2006	2007	2008	2009	2010	2011	2012	2013	Grand total
Bradford		1	2	2	24	158	375	396	163	108	1,229
Lycoming				5	11	23	119	300	202	163	823
Sullivan							22	19	27	14	82
Susquehanna			1	2	33	89	125	204	193	206	853
Tioga			1		15	123	276	272	122	32	841
Wyoming						2	24	70	15	67	178
Northern Tier total		1	4	9	83	395	941	1,261	722	590	4,006
All of Pennsylvania (37 counties)	*2*	*8*	*37*	*116*	*332*	*817*	*1,602*	*1,961*	*1,350*	*1,207*	*7,432*

Source: Pennsylvania Department of Environmental Protection

drilling activity. This may seem counterintuitive, because residents and others report very high and noticeable levels of industry activity in the counties. Yet high levels of activity by themselves do not guarantee a strong connection with the local economy. For example, if the workers are employed by a nonlocal company based outside of the county (or outside of Pennsylvania) and live in a neighboring county and only drive into the community for their shift and their work mostly relies upon supplies and equipment purchased outside of the county, their connection to the local economy may only be when they buy a sandwich or cup of coffee at a local restaurant or gas station. Heavy traffic and many workers passing through related to shale gas development do not necessarily mean a strong local economic impact any more than does having an interstate pass through a community guaranteeing major local economic benefits.

How much of the economic benefit of Marcellus Shale development stays within the counties where drilling is occurring is important to know because the communities with such activity are most directly bearing the costs of that development, including the nuisances and risks. Shale gas development creates some social and environmental challenges for host communities (see, e.g., Jacquet 2009; Brasier et al. 2011; Kragbo et al. 2010; Rozell and Reaven 2012; Roy 2013), so identifying the local economic benefits is important for understanding the implications of Marcellus Shale development in the communities where it occurs.

An Introduction to the Northern Tier

The six counties of Pennsylvania's Northern Tier are located in the northeast region of the state along the New York border. It is a rural part of the state with a population of almost 300,000 spread over 5,200 square miles. A significant amount of the landscape is forested, with much of the remainder being farmland. With a few exceptions, the major transportation networks within the region are two lane roads, which can make the region somewhat difficult or time consuming to traverse. It is relatively far from major urban and population centers. The total population within the Northern Tier has remained more or less stagnant for the past several decades (Fig. 2), reflecting generally struggling rural economies seeking ways to sustain and revitalize their communities. It is in this context that Marcellus Shale development is occurring, and which helps explain some of the positive local interest in the activity.

With the exception of parts of Sullivan County, there is no history of coal or gas extraction among the Northern Tier counties. That the region has such little history

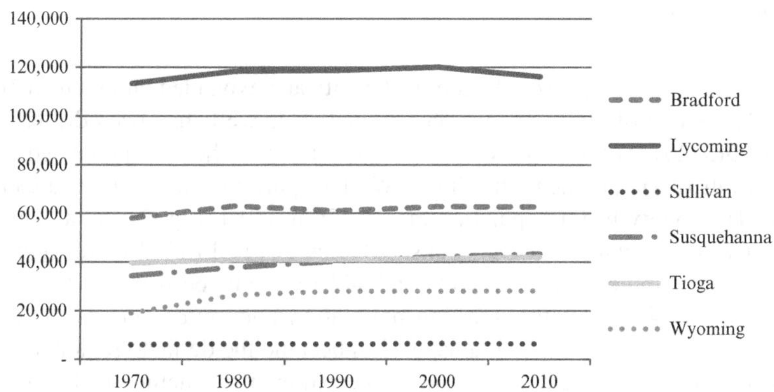

Fig. 2 Long-term population change in the Northern Tier (Source: US Census Bureau, State & County QuickFacts)

of extraction is important for two reasons. First, unlike other areas of the state with a history of gas or coal extraction (such as counties in southwest Pennsylvania), the Northern Tier did not have existing infrastructure to support the industry. Unlike many southwest counties, counties in the Northern Tier started with a clean slate, without many local businesses or residents with the experience or skills for natural gas drilling and development. The region is home to several interstate gas transmission pipelines, but similarly lacked other necessary gathering pipelines and other transportation infrastructures. Second, it is far more likely that land ownership and mineral right ownership are still intact in the Northern Tier counties (again, in contrast to counties in the southwest where these ownerships were more likely severed generations ago with the coal mining activity). This is noteworthy because ownership of mineral rights determines who receives lease and royalty dollars from natural gas companies. If land ownership and mineral right ownership have not been severed, so the surface owners also own the mineral rights, it means the surface owners where drilling activity occurs will receive those dollars, rather than others who are not so directly experiencing the physical consequences of that activity.

Bradford County

Bradford County leads the state in the number of shale wells drilled, with a total of 1,229 wells by 2013 (DEP 2014) (Table 1). Like most of the region, it is very rural with a population of 62,591 recorded by the 2010 Census (Fig. 2), and a population density of 54.6 persons per square mile (US Census Bureau 2014). In 2011 there were 1,535 business establishments in the county and 20,833 employed persons (BLS 2014). An estimated 23.9 % of the county's workforce commutes outside the county for work (US Census Bureau 2009) suggesting the county economy is more or less self-reliant.

Lycoming County

Shale drilling began in 2007 in Lycoming County and expanded rapidly in 2010 and 2011. By the end of 2013, 823 unconventional shale wells had been drilled in the county (DEP 2014) (Table 1). Lycoming County is by far the most populated county in the Northern Tier, home to the City of Williamsport, the largest urban area in the region. The county had a population of 116,176 in 2010 (Fig. 2) and a density of 94.6 persons per square mile (US Census Bureau 2014). In 2011 there were 3,016 business establishments in the county and 44,912 employed persons (BLS 2014). An estimated 14.5 % of the working population commutes outside the county (US Census Bureau 2014). Williamsport has become the de facto regional hub for gas development within the region due to its existing infrastructure, including housing (including hotels), regional airport, rail facilities, highway access, and available industrial sites for storage, maintenance, and repair of equipment and supplies. Anecdotally, many of the Marcellus Shale workers live in the Williamsport area due to housing availability and commute each day to the neighboring counties.

Sullivan County

Sullivan County is relatively new to Marcellus Shale drilling, despite its location in the heart of the Northern Tier. A lack of pipeline and other necessary infrastructures had delayed development until recently. The first well was drilled in 2010, and compared to the activity in some of the other Northern Tier counties, drilling activity in Sullivan County has remained slow. By 2013 the county had a total 82 shale wells (DEP 2014) (Table 1). Sullivan County is also the smallest county in the region in terms of population, with only 6,407 residents in 2010 (Fig. 2) and a density of only 14.2 people per square mile (US Census Bureau 2014). There were a total 169 business establishments in the county in 2011 and 1,321 employed persons (BLS 2014). An estimated 41.4 % of Sullivan County's workforce commutes outside of the county for work (US Census Bureau 2014), indicating that the local economy is very reliant on the surrounding outside economies.

Susquehanna County

Susquehanna County was one of the first Pennsylvania counties to experience Marcellus Shale activity, with its first well having been drilled in 2006. Between 2006 and 2013, there have been 853 wells drilled across the county (DEP 2014) (Table 1), making it the third most active county in the state in terms of wells drilled. It is a relatively rural county, with a population of 43,364 in 2010 (Fig. 2), a population density of 52.7 persons per square mile (US Census Bureau 2014). In 2011 there were 942 business establishments in the county, employing 7,159 people (BLS 2014). An estimated 49.3 % of the county's working population commutes outside of the county for work (US Census Bureau 2014).

Tioga County

Tioga County is the fourth most active Pennsylvania County in terms of number of wells drilled by 2013. Drilling activity in Tioga County began in 2008 and intensified rapidly through 2011; however in the past 2 years, drilling activity has slowed considerably and in 2013 just 32 wells were drilled (DEP 2014) (Table 1). By 2013 a total of 841 unconventional shale wells had been drilled in the county. Tioga County is also relatively rural, with a population of 42,025 in 2010 (Fig. 2), and a population density of 37.1 people per square mile (US Census Bureau 2014). In 2011 there were 993 local business establishments in the county and 11,072 employed persons (BLS 2014). An estimated 24.2 % of the county's working population work outside Tioga County (US Census Bureau 2014).

Wyoming County

Wyoming County has experienced much less drilling activity than some of the other Northern Tier counties because of its late start. The first well was drilled in the county in 2009, and a total of 178 wells have been drilled through 2013 (DEP 2014) (Table 1). However, because of its location along Route 6, a major east–west transportation corridor, it has been experiencing much other natural gas-related activities. It is a very small rural county, with a population of 28,257 in 2010 (Fig. 2), a population density of 71.1 persons per square mile (US Census Bureau 2014). In 2011 the county was home to 661 business establishments and has 8,580 employed persons (BLS 2014). About 47.3 % of the county's labor force works outside the county, indicating that the local economy is very dependent upon other outside economies (US Census Bureau 2014).

The Local Economic Impacts

The economic impact of Marcellus Shale development in these counties can be considered from several perspectives, including changes in local business activity, resident income, and employment. Each will be considered in turn.

Local Business Activity

The impact on local business activity is important to understand because dollars spent locally are more likely to stay local and recirculate in the community, creating greater indirect and induced economic benefits. Many of the industry's needs are highly specialized and were not available locally in the beginning of the play, such as drilling and fracking equipment, pipe, and sand. Some of the nonspecialized items, such as aggregate for well pad and road construction, food service, trucking, and construction, were available and are being purchased locally. In addition, the activity has led to new investment, such as hotel construction (Mount, Kelsey and Brasier, forthcoming).

One indicator of retail activity within the counties is state sales tax collections. Higher local retail sales mean more state sales tax collections, while declining local retail sales mean lower collections (though changes in sales tax collections don't perfectly track retail sales because food and clothing are excluded from the tax).

Table 2 Change in sales tax remittance in the Northern Tier

Adjusted for inflation, ($) in thousands					
			Percent change		Percent change
County (wells by 2011)	2007–2008 Remittance	2011–2012 Remittance	July 1, 2007, to June 30, 2012 (%)	2012–2013 Remittance	July 1, 2007, to June 30, 2013 (%)
Bradford (959)	$12,144	$18,929	55.9	$17,656	45.4
Lycoming (459)	$32,087	$35,613	11.0	$34,392	7.2
Sullivan (41)	$1,069	$1,330	24.4	$1,188	11.2
Susquehanna (454)	$8,022	$10,461	30.4	$10,849	35.2
Tioga (685)	$7,582	$8,444	11.4	$7,617	0.5
Wyoming (96)	$7,290	$8,409	15.4	$7,595	4.2
Northern Tier regional avg.	*$11,366*	*$13,864*	*24.7*	*$13,216*	*17.3*
Pennsylvania	*$8,496,554*	*$8,086,011*	*−4.8*	*$8,031,746*	*−5.5*

Source: PA Department of Revenue, Tax Compendium

The Pennsylvania Department of Revenue regularly releases data series on local county collections of the state sales tax in an annual tax compendium.

The data reveals that the Northern Tier counties experienced an average 24.7 % increase in sales tax collections between fiscal year 2007–2008 and 2011–2012 (Table 2). This is a very large increase when compared to the statewide average 4.8 % decrease at the county level during this time. Many counties experienced even larger increases in sales tax collections than the regional average. For example, Bradford County, the state leader in Marcellus Shale wells, experienced a 55.9 % increase during this time period and Susquehanna County experienced an increase of 30.4 %.

The data suggest retail activity is declining from its high in 2011 to 2012; between 2011–2012 and 2012–2013, sales tax collections fell in all but one of the Northern Tier counties (Table 2). This is likely due to the decrease in well activity in the state in 2012 and 2013 as gas prices dropped as well as a reflection of the phases of development within the natural gas industry. From fiscal years 2007–2008 to 2012–2013, sales tax collections increased an average 17.3 % in the Northern Tier compared to a 5.5 % decrease at the state level. Collections in both Bradford and Susquehanna counties were once again the largest (45.4 % increase and 35.2 % increase, respectively).

Net Profits

Another way to measure local economic activity is through local business profits. The number of tax returns from Northern Tier counties reporting net profits income, which is paid by local business owners on the profits of their business activity, declined an average 2.8 % between 2007 and 2011 compared to an average 1.5 %

Table 3 Change in net profits income in the Northern Tier, 2007–2011

Adjusted for Inflation, ($) in thousands						
County (wells by 2011)	Number of returns		Percent change	Income		Percent change
	2007	2011		2007	2011	
Bradford (959)	3,711	3,532	−4.8	$72,042	$123,834	71.9
Lycoming (459)	5,952	5,790	−2.7	$163,965	$187,382	14.3
Sullivan (41)	449	458	2.0	$9,267	$10,039	8.3
Susquehanna (454)	2,981	2,856	−4.2	$59,378	$82,920	39.2
Tioga (685)	2,377	2,484	4.5	$43,488	$78,647	80.8
Wyoming (96)	1,959	1,728	−11.8	$44,837	$59,658	35.8
Northern Tier regional avg.	*2,905*	*2808*	*−2.8*	*$65,530*	*$90,619*	*41.7*
Pennsylvania	*680,322*	*690,843*	*1.5*	*$26,952,540*	*$26,936,318*	*−0.1*

Source: PA Department of Revenue PIT Statistics

increase across the state (Table 3). This suggests that the number of locally owned businesses declined during this period as a result of the shale activity.

However, the net profit income reported by the declining number of businesses increased dramatically in several of the counties. In Bradford County, despite a 4.8 % decline in the number of tax returns reporting net profits income, there was a 71.9 % increase in gross profit income reported. Similar changes were seen in Susquehanna and Wyoming counties. Across all six counties, the average increase in net profits income was 41.7 % compared to a 0.1 % average decline across all of Pennsylvania.

The personal tax income data suggests that the natural gas activity has had a negative impact on the number of locally owned businesses, but that surviving locally owned businesses on average experienced large increases in profits. Anecdotes from local residents and business owners reveal that there has been increased competition between local businesses and nonlocal firms as a result of the gas development, as new firms follow the activity into the counties. In addition, the decline in the number of taxpayers reporting net profits could result from local businesses buying up local competitors and consolidating, or from local business owners closing their business in order to take advantage of new work opportunities with the natural gas companies.

Local Business Owners' Perceptions

A study conducted by Kelsey et al. (2011) surveyed local businesses in Bradford County and Washington County (southwest Pennsylvania) to determine local business owners' perceptions of impacts related to the Marcellus Shale development. The survey results provide a look at the impacts related to Marcellus in greater detail by distinguishing impacts by type of business.

Table 4 Local business owners' perceptions of Marcellus-related impacts

Results reflect perceptions of 360 survey respondents from Bradford County		
	Percent (number) saying yes	
Business type	Have your business activities changed due to natural gas drilling?	Have your annual sales increased due to natural gas drilling?
Agriculture, forestry, fishing	9 % (2)	9 % (2)
Mining	–	–
Construction	35 % (8)	27 % (6)
Manufacturing	11 % (3)	25 % (7)
Transportation, communications, utilities	30 % (3)	22 % (2)
Wholesale trade	28 % (5)	33 % (6)
Retail trade	25 % (13)	44 % (23)
Financial, insurance, real estate	28 % (7)	50 % (12)
Business services	20 % (10)	33 % (16)
Professional services	15 % (9)	23 % (13)
Eating and drinking places	29 % (6)	38 % (8)
Hotels and campgrounds	80 % (4)	100 % (5)

Source: Kelsey, Shields, Ladlee, and Ward (2011)

Several business types appear to have experienced more positive impacts than others (Table 4). Their survey results indicate that 27 % of construction businesses have noticed an increase in annual sales as a result of the natural gas drilling. Similarly 25 % of manufacturing and 22 % of transportation, communications, and utilities companies noticed an increase in annual sales. An even greater percentage of companies within retail trade, eating and drinking places, and within finance, insurance, and real estate noticed that natural gas drilling has positively impacted their annual sales (44 %, 38 %, and 50 %, respectively). One hundred percent of the surveyed hotels and campgrounds businesses said they had experienced an increase in their annual sales because of natural gas drilling. This is likely due to the large portion of workers within the industry who have relocated only temporarily to the county for work and therefore require only temporary housing accommodations.

Personal Income

Resident personal income similarly has been impacted greatly by the natural gas activity in the region. The changes in personal income can be observed by looking at data collected by the Pennsylvania Department of Revenue. Their data, which distinguishes changes by type of income, allows interpreting more precisely how local residents' total personal income has changed and the portion of residents affected. The three largest contributors of personal income are gross compensation (i.e., wages and salaries), net profits (discussed earlier), and rents, royalties, patents, and copyrights income.

Total Taxable Income

Total taxable income increased an average 11.8 % across the Northern Tier in the years between 2007 and 2011 (Table 5). During this time, total taxable income decreased an average 7.6 % at the county level across all of Pennsylvania. The number of total taxable income tax returns filed in the Northern Tier counties decreased an average 0.5 % at the county level, indicating there were fewer taxpayers reporting income. Though this was a decline, it was less than the 1.5 % average decrease at the county level statewide during the same time period.

The taxable income increases in several of the counties were extremely large, particularly in comparison to the statewide decline. Residents of both Bradford and Tioga counties reported a 25.4 % increase in taxable income during this time period, while residents of Susquehanna reported a 12.7 % increase. The income increases were much larger than the increase in the number of tax returns filed, indicating that the majority of such income increases result from higher income going to residents rather than simply a growth in the population.

Gross Compensation

Gross compensation income represents wages and salaries earned by county residents. As seen in Table 5, gross compensation income reported by Northern Tier county residents increased an average 4.1 % between 2007 and 2011, and the number of tax returns reporting this type of income increased in 1.3 %. In contrast, across Pennsylvania counties, gross compensation income decreased an average 2.3 % and the number of returns decreased an average 0.6 %. Experiences of the individual

Table 5 Change in personal income, 2007–2011

Income adjusted for inflation								
	Total taxable income		Gross compensation		Net profits		Rents, royalties, patents, and copyrights	
County (wells by 2011)	Returns (%)	Income (%)	Returns (%)	Income (%)	Returns (%)	Income (%)	Returns (%)	Income (%)
Bradford (959)	2.6	25.4	4.5	9.0	−4.8	71.9	88.8	960.5
Lycoming (459)	−2.0	1.0	−0.1	1.8	−2.7	14.3	39.9	184.0
Sullivan (41)	−0.5	7.9	3.0	4.8	2.0	8.3	138.1	598.4
Susquehanna (454)	1.4	12.7	2.3	2.0	−4.2	39.2	45.4	433.2
Tioga (685)	4.2	25.4	6.3	15.1	4.5	80.8	78.2	636.9
Wyoming (96)	−8.8	−1.7	−8.1	−8.2	−11.8	35.8	42.7	239.5
Northern Tier regional avg.	−0.5	11.8	1.3	4.1	−2.8	41.7	72.2	508.8
Pennsylvania	−1.5	−7.6	−0.6	−2.3	1.5	−0.1	15.7	37.1

Source: PA Department of Revenue, PIT Statistics

Table 6 Change in gross compensation income in the Northern Tier, 2007–2011

Adjusted for inflation, ($) in thousands						
County (wells by 2011)	Number of returns		Percent change	Income		Percent change
	2007	2011		2007	2011	
Bradford (959)	21,509	22,471	4.5	$808,113	$881,115	9.0
Lycoming (459)	44,082	44,017	−0.1	$1,682,331	$1,712,589	1.8
Sullivan (41)	2,100	2,164	3.0	$70,625	$73,987	4.8
Susquehanna (454)	14,214	14,536	2.3	$511,192	$521,562	2.0
Tioga (685)	13,673	14,540	6.3	$473,436	$544,772	15.1
Wyoming (96)	11,959	10,995	−8.1	$453,485	$416,493	−8.2
Northern Tier regional avg.	*17,923*	*18,121*	*1.3*	*$666,530*	*$691,753*	*4.1*
Pennsylvania	*4,654,462*	*4,624,863*	*−0.6*	*$232,680,601*	*$227,396,476*	*−2.3*

Source: PA Department of Revenue, PIT Statistics

counties varied quite a bit, but with the exception of Wyoming County, all experienced more positive changes both in number of returns and income reported than the statewide average.

The change observed in some Northern Tier counties was even greater. In Bradford County the number of tax returns reporting gross compensation income increased in 4.5 % between 2007 and 2011, while in Tioga County they increased in 6.3 %. Because the tax data include a mix of single- and joint-tax returns, these increases may not necessarily perfectly match actual employment changes in the counties (Table 6).

Gross compensation income reported on resident tax returns increased at a greater rate than the increase in number of returns. Across the Northern Tier, gross compensation income reported increased an average 4.1 % between 2007 and 2011, and in three of the individual counties, the increase was even greater. Gross compensation income reported on tax returns increased an average 9 % in Bradford County, 4.8 % in Sullivan County, and 15.1 % in Tioga County.

The tax return data shows that there has been a small yet noticeable increase in local resident employment and a substantial increase in taxable wage and salary income. This suggests that the greatest employment impact of Marcellus activity for county residents has been an increase in pay or an increase in hours worked (or both) rather than an increase in employment opportunities.

Rents, Royalties, Patents, and Copyrights Income

Another major source of income to residents from the shale gas activity is lease and royalty payments from gas companies to mineral right owners in exchange for being able to access and extract the gas. Between 2007 and 2011, the number of tax returns from county residents reporting rents, royalties, patents, and copyrights income increased an average 72.2 % in the Northern Tier, compared to a 15.7 % increase

Table 7 Change in rents, royalties, patents, and copyrights income in the Northern Tier, 2007–2011

Adjusted for inflation, ($) in thousands						
	Number of returns		Percent	Income		Percent
County (wells by 2011)	2007	2011	change	2007	2011	change
Bradford (959)	2,261	4,269	88.8	$14,547	$154,269	960.5
Lycoming (459)	2,596	3,633	39.9	$3,633	$81,755	184.0
Sullivan (41)	197	469	138.1	$1,315	$9,185	598.4
Susquehanna (454)	1,514	2,201	45.4	$14,406	$76,809	433.2
Tioga (685)	1,354	2,413	78.2	$9,340	$68,830	636.9
Wyoming (96)	910	1,299	42.7	$9,226	$31,321	239.5
Northern Tier regional avg.	1,472	2,381	72.2	$8,745	$70,361	508.8
Pennsylvania	234,918	271,834	15.7	$3,342,823	$4,584,546	37.1

Source: PA Department of Revenue, PIT Statistics

statewide (Table 7). The increase in the number of tax returns from taxpayers reporting this type of income was as great as 88.8 % in Bradford County and 138.1 % in Lycoming County. Such large increases reflect a growing number of county residents receiving this type of income.

The income changes reported on these tax returns are even more significant. In 2007, residents of the six Northern Tier counties reported receiving a total of $52.5 million in rents, royalties, patents, and copyrights income (Table 7). In 2011, 5 years later, Northern Tier county residents reported receiving $422 million in rents, royalties, patents, and copyrights income, a 508.8 % increase in reported income of this type. Statewide, reported income of this type increased just 37.1 % during this time period.

As with the net profit and gross compensation income, the experience of individual counties varied quite a bit from these regional averages. Residents of Bradford County reported a 960.5 % increase in such income, while the change in Tioga and Sullivan was 636.9 % and 598.4 %, respectively. Only in Lycoming and Wyoming were the increases significantly below the regional average (184 % and 239.5 %, respectively).

Unfortunately, available data on lease and royalty income is only as recent as 2011, at the height of drilling activity in most counties across the state and the Northern Tier. Since 2011, drilling activity has decreased considerably in many counties due to falling gas prices and a new concentration in drilling in other states with shale. Because lease and royalty payments are directly related to drilling activity, it is very likely that lease and royalty income has also decreased considerably in the last 2 years.

Composition and Distribution of Income

The tax return data indicates there were fairly large changes in resident income between 2007 and 2011. It is important to consider the overall composition of these income changes, because they directly affect the distribution of the economic benefits of the gas development activity among residents.

Table 8 Real changes in personal income, 2007–2011

Adjusted for inflation (1,000's)

County (wells by 2011)	Annual gross compensation income		Annual net profits income		Annual rents, royalties, patents, and copyrights income	
	Percent of returns	Total income change	Percent of returns	Total income change	Percent of returns	Total income change
Bradford (959)	82.0	$73,002.24	12.9	$51,792.26	15.6	$139,722.25
Lycoming (459)	84.3	$30,257.56	11.1	$23,416.99	7.0	$52,966.54
Sullivan (41)	75.9	$3,361.78	16.1	$772.00	16.5	$7,869.52
Susquehanna (454)	79.8	$10,370.30	15.7	$23,341.66	12.1	$62,403.24
Tioga (685)	81.0	$71,335.54	13.8	$35,159.25	13.4	$59,490.40
Wyoming (96)	82.7	–$36,991.57	13.0	$16,056.04	9.8	$22,094.82
Northern Tier regional total	81.0	$151,335.86	13.8	$150,538.19	12.4	$344,546.77

Source: PA Department of Revenue, PIT Statistics

The tax return data indicates that rents, royalties, patents, and copyrights income to residents of Northern Tier counties increased more in real terms than did the other sources of resident income (Table 8). Across the six counties, for example, such annual income reported by residents increased by $344.5 million between 2007 and 2011, more than twice the increase in gross compensation ($151.3 million) and increase in net profits ($150.5 million) (all numbers adjusted for inflation). These calculations are the changes between incomes reported in 2007 and in 2011, not the cumulative change income across these years, which would be much higher. The data indicates that lease and royalty payments to the owners of mineral rights thus are the largest positive economic impact from the shale development in the counties.

The difference between gross compensation income and rents, royalties, patents, and copyrights income was very large in some counties. In Susquehanna County, for example, taxpayers reported an annual increase of $62.4 million in lease and royalty income, six times more than the $10.4 million annual increase they reported in gross compensation. Only in Tioga County did increases in gross compensation exceed increases in rents, royalties, patents, and copyrights income.

The tax return data clearly indicate that the lease and royalty income went to a relatively small share of local residents. In Bradford County, for example, 15.6 % of residents' tax returns reported such income (see Table 9), while only 7 % of tax returns from Lycoming County did so. This means only a relatively small proportion of residents are receiving the largest local economic benefits from the shale activity. Other local economic benefits, such as the number of jobs created, the increase in wages received, or increase in local business profits, are important and do contribute

Table 9 Distribution of lease and royalty income in the Northern Tier

County (wells by 2011)	Number of returns, 2007		2007	Number of returns, 2011		2011
	Total taxable income	Lease and royalty income	Proportion of returns lease and royalty (%)	Total taxable income	Lease and royalty income	Proportion of returns lease and royalty (%)
Bradford (959)	26,705	2,261	8.5	27,404	4,269	15.6
Lycoming (459)	53,289	2,596	4.9	52,217	3,633	7.0
Sullivan (41)	2,864	197	6.9	2,851	469	16.5
Susquehanna (454)	17,963	1,514	8.4	18,208	2,201	12.1
Tioga (685)	17,220	1,354	7.9	17,945	2,413	13.4
Wyoming (96)	14,585	910	6.2	13,297	1,299	9.8
Northern Tier regional avg.	22,104	1,472	7.1	21,987	2,381	12.4
Pennsylvania	5,614,665	234,918	4.2	5,527,878	271,834	4.9

Source: PA Department of Revenue, PIT Statistics

to the local economies, yet they pale in comparison to the size of the lease and royalty checks some residents of these counties are receiving. At the same time, the hardships and challenges that are created by the shale development, such as the increased road traffic, typically affect all residents.

There are numerous factors which may contribute to why the portion of residents receiving lease and royalty dollars is so small. First, only about half the land (51 %) with Marcellus Shale in Pennsylvania is owned by in-county residents (Kelsey et al. 2011). Much of the land owned by nonresidents are vacation properties, such as second homes or hunting camps. In addition, a large portion of the land is owned by the Commonwealth itself. Any lease and royalty dollars paid to out-of-county residents or to the Commonwealth immediately leave the communities in which the drilling is occurring. Second, many residents in the counties with drilling do not own property and instead rent their residences. They therefore have no mineral rights to lease to the natural gas companies.

Land ownership records suggest that even among those receiving lease and royalty income, the payments will be highly concentrated among a relatively small share of landowners. Kelsey et al. (2012) used landownership data from planning offices in eleven Pennsylvania counties with Marcellus Shale development and found that the majority of landowners in these counties own relatively small parcels; in Bradford County, for example, 38.6 % of the resident landowners owned less than an acre of land, and half owned less two acres. They sorted landowners by the total acreage owned and found that 80 % of the local landowners in Bradford County together own 3.7 % of the county's land area, while the top 10 % of landowners together own 43.9 %. Lease and royalty dollars generally will follow the same distribution pattern, which suggests that the majority of these economic benefits are going to a relatively small share of county residents.

Employment

A workforce needs study of shale gas development in Pennsylvania found that the employment needs directly related to drilling are very broad, including almost 150 occupations (Brundage et al. 2011). In addition, relatively few of these jobs are highly specialized to the natural gas industry; for example, general office staff and laborers both account for 20 % of the workforce required to drill one Marcellus well, while heavy equipment operators and truck drivers account for an additional 17 % and 10 %, respectively (see Fig. 3). If indirect employment needs are included, such as food and accommodation service jobs, the potential employment opportunities are even broader.

A stereotype often heard in the shale counties is that the jobs largely have gone to nonresidents and that few Pennsylvania residents possess the specialized skills required for much of the work directly with the natural gas companies. There has been a growth of shale-related workforce training programs in Pennsylvania since the onset of Marcellus Shale development, such as at Lackawanna College, Penn College, Westmoreland Community College, and others.

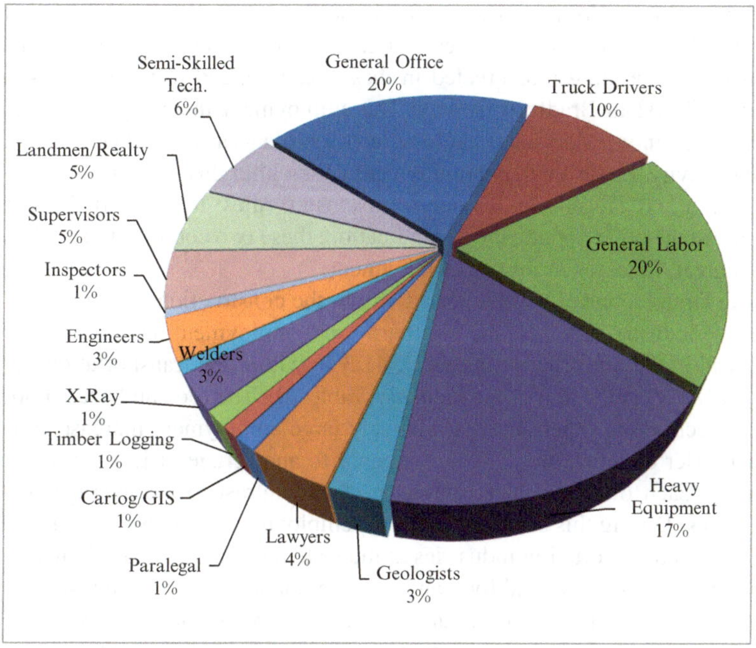

Fig. 3 Workforce required to drill one Marcellus Shale Well

Employment by Industry

The US Bureau of Economic Analysis (BEA) regularly releases data on employment at the county level which breaks down employment by major NAICS industry. An analysis of this data set allows understanding more precisely how employment has been affected by the Marcellus Shale activity. This federal data is reported by employers, reflecting how many people they have working in the specific county without regard to where the workers live. In other words, the employment numbers include nonresidents commuting into or living temporarily in the county and so do not necessarily reflect employment by residents.

Table 10 shows the total change in employment in the Northern Tier counties, as well as changes in employment in six major industries. These six industries (mining, construction, retail, transportation, real estate, and food and accommodation services) are the most likely to be affected by the natural gas extraction activity as they relate directly to the shale activity, supporting industries, or tourism. Total employment, as reported by the BEA, increased an average 4 % in the Northern Tier counties, but decreased an average 0.7 % at the state level. These numbers are not all that different from the employment change depicted by the PA Department of Revenue data. The Northern Tier regional average changes were more positive than the average changes reported statewide within each of the major industries, strongly suggesting that the Marcellus activity has had an important impact on employment.

Not surprisingly, the mining industry experienced the largest increase in employment in the Northern Tier counties, increasing on average 166.4 % between 2007 and 2011, compared to an average 62.7 % increase across the state (Table 10). The number of mining jobs created in these years ranged from 206 in Wyoming County to 1,502 in Bradford County. The employment increase of 1,054.3 % in Lycoming County (1,455 jobs) likely reflects its role as a regional center for development activity. Unfortunately employment data within this industry is unavailable for Sullivan and Tioga counties due to disclosure rules; however if the increases observed in the other counties is any indication, these two counties likely also experienced great increases in mining employment.

Other large increases were reported within the construction, transportation, and real estate industries. Across the Northern Tier, employment increased an average 6.7 % within the construction industry, 41.3 % within the transportation industry, and 24.9 % within the real estate industry (Table 10). The food and accommodation services sector also experienced a relatively large employment increase across the Northern Tier counties of 11.1 %, compared to an average of just 1.1 % increase statewide. Each of these changes was in stark contrast to the average statewide experiences. During this same time period, employment decreased within the construction and transportation industries at the state level and increased only marginally within the real estate and food and accommodations services industries.

The observed employment changes within the retail industry are interesting, given the average relatively large increases in sales tax activity occurring in these counties. Across the Northern Tier, there was an average 2.8 % decline in employment

Table 10 Change in employment in major industries, 2007–2011

County (wells by 2011)	Total employment	Mining	Construction	Retail	Transportation	Real estate	Food and accommodations
Bradford (959)	2,669 (8.4 %)	1,502 (315.5 %)	219 (13.3 %)	94 (2.4 %)	629 (52.6 %)	337 (47.6 %)	254 (18.3 %)
Lycoming (459)	1,459 (2.1 %)	1,455 (1054.3 %)	290 (8 %)	−564 (−6.5 %)	(D)	61 (3.7 %)	403 (10.3 %)
Sullivan (41)	−45 (−1.5 %)	(D)	0 (0 %)	−54 (−18.2 %)	(D)	(D)	−6 (−3.6 %)
Susquehanna (454)	569 (3.4 %)	655 (41.2 %)	−48 (−3.4 %)	−211 (−10.3 %)	(D)	163 (32.5 %)	31 (3.4 %)
Tioga (685)	1,000 (5.3 %)	(D)	194 (22.9 %)	13 (0.5 %)	200 (32.9 %)	239 (49.3 %)	160 (12.9 %)
Wyoming (96)	404 (3.3 %)	206 (228.9 %)	−78 (−8.9 %)	190 (14.1 %)	230 (30.2 %)	82 (38.9 %)	75 (12 %)
Northern Tier regional average	1,009 (4 %)	955 (166.4 %)	96 (6.7 %)	−89 (−2.8 %)	353 (41.3 %)	176 (24.9 %)	153 (11.1 %)
Pennsylvania	−48,660 (−0.7 %)	20,453 (62.7 %)	−56,315 (−13.6 %)	−39,141 (−4.9 %)	−3,843 (−1.4 %)	14,988 (6 %)	4,892 (1.1 %)

Source: BEA CA25N, Total Full-Time and Part-Time Employment
(D) Data not available to due disclosure policy

within this sector, despite the very large increases in retail sales activity. The region did slightly better than the average 4.9 % decrease statewide, yet not as much as would be expected given the increased sales tax collections. In three of the counties (Lycoming, Sullivan, and Susquehanna), the average decreases in employment within the retail industry were even greater than the statewide average decline (6.5 % decrease, 18.2 % decrease, and 10.3 % decrease, respectively). The difference between retail activity (as reflected by sales tax collections) and employment indicates that the higher retail activity is not resulting in much job creation in the region.

The data suggests there have been major impacts on employment as a result of the Marcellus activity. Large increases in employment within several major industries, namely, the mining and transportation sectors, likely have benefitted both local residents and commuter workers. The Pennsylvania Department of Revenue data suggests that county resident workers are benefitting not only from the increase in number of jobs available but from an increase in earned wages too. Comparing the Department of Revenue tax return data and the BEA employment data shows that local employment was more positively impacted in several of the individual counties. The data for Sullivan County, for example, suggests an overall 1.5 % decrease in total employment, but a 3 % increase in the number of tax returns, reflecting local workers earning wages. Tioga County similarly appears to have experienced a more positive impact on specifically local employment. Other counties, including Bradford, Lycoming, and Wyoming, experienced much more positive changes in total employment reported by the BEA than changes in employment suggested by the Department of Revenue tax return data, which would suggest that much of the employment changes there are due to nonresidents.

Conclusion

The various economic analyses, surveys, and accounts of local perceptions of change paint a pretty clear picture of the local economic impact in the Northern Tier as a result of the Marcellus Shale development. On a variety of economic measures, such as changes in local business numbers, employment, and resident income, the Northern Tier counties with Marcellus Shale activity are doing much better than the statewide trends. Changes in the counties with the most drilling activity, such as Bradford and Tioga, particularly stand out. Many local businesses noticed an increase in sales and business profits as a result of the Marcellus drilling. Especially affected were businesses in the accommodations, real estate, and retail sectors of the economy. Local collections of Pennsylvania state sales tax confirm that retail activity in these communities has increased dramatically since 2007 – as high as 45 % (in Bradford County). These experiences in the Northern Tier counties differ greatly from the average experience of declining sales and business profits statewide.

County resident income also appears to have been positively affected by the drilling activity in the region, as evidenced by the average 11.8 % increase in total taxable income in the Northern Tier and especially the 25.4 % increases in resident taxable income reported in Bradford and Tioga counties. Such increases are extremely large in such a short time span, particularly when the average county-level change in the state was a decline of 7.6 %. Resident income increased in these counties due to increases in wages and salaries, increases in business profits, and, above all, increases in lease and royalty payments to mineral right owners. County resident tax returns showed repeatedly that each of the types of income (gross compensation, net profits, and rents, royalties, patents, and copyrights) on average increased at a greater rate within the Northern Tier than across the entire Commonwealth.

Employment data from employers suggests stronger job creation in the Northern Tier between 2007 and 2011, with most of these new jobs occurring within the mining and transportation industries. Yet state tax returns suggest that many of these new jobs went to noncounty residents, rather than to people who live in the counties. In addition, the data indicate employment was affected more by increased wages to the existing workforce than by an overall increase in the number of jobs. Across the Northern Tier, gross compensation income reported on tax returns increased an average 4.1 % compared to a 1.3 % increase in the number of tax returns filed.

This chapter has illustrated many of the local economic changes being experienced in the counties of Pennsylvania's Northern Tier, yet it is difficult to fully understand how local residents and businesses are being impacted without thoroughly considering the local costs of the activity as well. It is difficult at this stage of the development to quantify environmental and other costs arising due to Marcellus Shale development in the region, but such costs are an equal part of the experiences of these counties even if they currently have not been adequately measured. There is widespread concern, for example, about the impact of the drilling activity on tourism in the region. The overall economic impact in the counties includes both benefits and costs, yet only the benefits can currently be quantified, so such benefits must be considered in the context of uncertainty over costs.

It is also important to keep in mind when evaluating these changes that the individual experiences of residents, of businesses, and of communities in these counties, so the averages do not represent everyone's experiences. The changes described in this chapter are averages which illustrate general trends in local employment and income and may not reflect everything that is occurring within each of the counties. However comparing the regional and county-level changes to the statewide average changes does provide some perspective and perhaps lend understanding to what might have occurred in the Northern Tier in the absence of Marcellus Shale development.

The changes described in this chapter must be considered short-run impacts of the shale development, reflecting immediate changes in employment, business growth, and personal income as a result of the drilling phase of Marcellus Shale development. Drilling in the region peaked in 2011, the most recent year that Pennsylvania Department of Revenue data is available, and by 2013 had dropped more than 50 % in the region due to changes in natural gas prices. What this portends for the future of these counties is unclear; is this a brief lull in a drilling "boom," or is this the beginning of the "bust"? Will drilling continue at a slower but steadier pace? No one really knows yet. It could be argued that a slower pace of development would be beneficial for economic development in the counties, because it would lengthen the amount of time the activity will occur in the region. In addition, a slower pace with fewer workers would be less likely to overwhelm these relatively small economies, with less pressure on housing and other infrastructure and more time for local workers and businesses to adapt and compete rather than having to rely as heavily upon outside entities to fulfill labor and business requirements for development.

Regardless of the pace of drilling, this is a finite, nonrenewable resource which at some time will become depleted. What happens then is a critical question for these communities. Past experience with resource-based economic activity and the termed "boom-bust cycle" would suggest that the communities of the Northern Tier focus on maintaining a diverse economy in preparation for when the resource and the related activity are gone. The gas development will not be a permanent addition to the local economies of the Northern Tier, but rather must be viewed as a temporary infusion of economic activity. Whether and how the counties use this to build stronger, more sustainable economies in the long run remains to be seen.

References

Brasier KJ, Filteau MR, McLaughlin DK, Jacquet J, Stedman RC, Kelsey TW, Goetz SJ (2011) Residents' perceptions of community and environmental impacts from development of natural gas in the Marcellus Shale: a comparison of Pennsylvania and New York cases. J Rural Soc Sci 26(1):32–61

Brundage TL, Jacquet J, Kelsey TW, Ladlee JR, Lobdell J, Lorson JF, Michael LL, Murphy TB (2011) Pennsylvania statewide Marcellus Shale workforce needs. Marcellus Shale Education and Training Center, Williamsport

Jacquet J (2009) Energy boomtowns and natural gas: implications for Marcellus Shale local governments and rural communities, Rural development paper, No. 43. Northeast Regional Center for Rural Development, State College

Kargbo DM, Wilhelm RG, Campbell DJ (2010) Natural gas plays in the Marcellus Shale: challenges and potential opportunities. Environ Sci Technol 44:5679–5684

Kelsey TW, Metcalf A, Salcedo R (2012) Marcellus Shale: land ownership, local voice, and the distribution of lease and royalty dollars. The Center for Economic and Community Development, University Park

Kelsey TW, Shields M, Ladlee JR, Ward M (2011) Economic impacts of Marcellus Shale in Pennsylvania: employment and income in 2009. Marcellus Shale Education and Training Center

Mount DJ, Kelsey TW, Brasier KJ (forthcoming) The impact of Marcellus Shale development on hotel revenues in Pennsylvania. ICHRIE Penn State Research Reports

Pennsylvania Department of Environmental Protection (2014) SPUD data report. Pennsylvania Department of Environmental Protection, Harrisburg

Pennsylvania Department of Revenue (2008–2012). Tax compendium. 2007–08 through 2011–2012. Pennsylvania Department of Revenue, Harrisburg

Pennsylvania Department of Revenue (2008–2012) Personal income statistics. 2007, 2008, 2009, and 2010. Pennsylvania Department of Revenue, Harrisburg

Roy AA, Adams PJ, Robinson AL (2013) Air pollutant emissions from the development, production and processing of Marcellus Shale natural gas. J Air Waste Manage Assoc 64(1):19–37

Rozell DJ, Reaven SJ (2012) Water pollution risk associated with natural gas extraction from the Marcellus Shale. Risk Anal 32(8):1382–1393

Census Bureau US (2014) State & County QuickFacts. U.S. Department of Commerce, Washington, DC

U.S. Census Bureau (2005–2014) Commuting characteristics by sex: 2005–2009 American Community Survey 5-Year Estimates. U.S. Department of Commerce, Washington, DC

U.S. Bureau of Economic Analysis (2012) Local areas personal income and employment, CA25N Total Full-Time and Part-Time Employment by NAICS Industry. Accessed 28 June 2013

U.S. Bureau of Labor Statistics (2014) Quarterly census of employment and wages. U.S. Department of Labor, Washington, DC

Marcellus Shale and the Commonweal of Pennsylvania

Timothy W. Kelsey and Kirsten Hardy

Abstract Much of the enthusiasm about Marcellus Shale has been its promise of economic benefits. State and federal data suggests that Marcellus Shale is having generally positive but modest effects on employment, wages, and local business activity. Wages and income generally have increased more than the number of workers, suggesting that much of the impact has been more work hours, higher pay, or a combination of both, rather than significant new job creation. Employment is up, particularly in sectors directly related to drilling activity, yet the number of residents reporting wages and salaries has not changed as much, indicating that many of the new jobs are going to nonresidents. Counties with Marcellus Shale activity typically did a little better in retaining or adding local businesses than did the rest of the state. Many of these economic numbers appear more modest than would be expected, given the billions of dollars being spent to develop the Marcellus Shale.

Introduction

Much of the enthusiasm about Marcellus Shale development within Pennsylvania is focused upon the potential economic impacts of the activity. Anecdotes and news stories have noted new hiring, expanding businesses, large lease and royalty checks going to landowners, and workers moving into the Marcellus region. The amount of

T.W. Kelsey (✉)
Department of Agricultural Economics, Sociology and Education,
The Pennsylvania State University, University Park, USA
e-mail: tkelsey@psu.edu

K. Hardy
Former student in Community, Environment and Development,
The Pennsylvania State University, University Park, USA
e-mail: kirstenhardy81@gmail.com

W.E. Hefley, Y. Wang (eds.), *Economics of Unconventional Shale Gas Development*, Natural Resource Management and Policy 45,
DOI 10.1007/978-3-319-11499-6_5

.oney being spent to develop Marcellus in Pennsylvania is significant. Considine et al. (2011) reported that gas companies spent nearly $11.5 billion in Pennsylvania during 2010, with per well spending of around $5 million (Marcellus Shale Coalition 2011).

Marcellus Shale activity largely is occurring within rural counties that have struggled for decades with population loss and declining manufacturing and other employment. In addition, the onset of activity occurred simultaneously with the Great Recession, at a time when the business community and state government was looking for relief from the global economic downturn and plunging state tax revenues. Conversely, those same rural counties are also home to significant natural resources, including the largest contiguous hardwood forests east of the Mississippi and important headwaters to the Susquehanna and other major rivers, that some view as threatened by the development.

Given the wide attention, it should not be surprising that there has been disagreement and potentially competing studies of the potential statewide economic impacts (see, e.g., Considine et al. 2009; Herzenberg 2011; Kelsey et al. 2011; Kinnaman 2011; Politics 2011; Republican Party of Pennsylvania 2011; StateImpact 2013; Weinstein and Partridge 2011). These studies have focused on the short-term economic benefits of Marcellus Shale activity and largely have not addressed the possible long-term economic impacts of Marcellus Shale activity in Pennsylvania (in addition, none have adequately addressed the short-run costs of such activity).

There is evidence from prior energy booms that any positive employment impacts are fleeting and that after development ends, these economies are worse off than if the boom had never occurred (see Headwaters Economics 2008; Jacobsen and Parker 2014; Papyrakis and Gerlagh 2007; James and Aadland 2011). Some other studies have found conflicting results, though the local impacts are modest (Brown 2014).

At a minimum, such experience of past resource booms would suggest that Pennsylvanians should view the economic benefits from Marcellus Shale as a temporary phenomenon, regardless of the current size of those benefits. Even in the short years that Pennsylvania has experienced Marcellus Shale development activity, it could be argued that the Commonwealth has already experienced such a mini boom/bust cycle. Well drilling peaked in 2011, with 1,961 wells drilled that year, and then dropped to 1,207 in 2013 (a 38.4 % decline (PA DEP)) due to the falling natural gas prices and an industry shift to the Utica Shale in Ohio. Though the activity remains in Pennsylvania counties, it is much less apparent than during 2011.

This chapter examines the short-term statewide economic experience in Pennsylvania with Marcellus Shale development, using multiple federal and state data sets to gain a comprehensive view of that experience so far. The data includes federal data sources on wages and salaries, employment, and number of business establishments and state income and sales tax information. Most of the data currently available is through 2011, though several of the agencies have released 2012 and 2013 data, allowing those to be used. It is important to remember that 2011 was the peak of Marcellus Shale drilling activity in Pennsylvania, so the impacts today would be less than implied by the 2011 data.

Marcellus Shale Activity

Shale gas development requires a high level of specialization of businesses, equipment, and tasks, different than past gas development in the Commonwealth. Work crews and equipment generally are highly specialized, with each crew typically performing only a narrow range of the total work required on a site. These can include separate crews focused on geological studies, seismic testing, leasing, permitting, well pad development, water management, pipeline construction, compressor construction, drilling, fracking and completion, and reclamation (Brundage et al. 2011). The work crews typically shift frequently between work locations across multiple counties to perform their work, rather than being on the same site for a long time.

Many of the businesses involved in the development are regional, national, or multinational companies, with little formal footprint in the individual counties where drilling is occurring. Much of the equipment and supplies are highly specialized, such as the drilling rigs, drilling pipe, and frac sand, unavailable from county-based businesses (and in some cases, unavailable from within Pennsylvania). The result is that much of the gas industry spending on Marcellus Shale development occurs across Pennsylvania and in other states, not just in the locations with drilling activity.

Natural gas development by its nature has a limited time span because it is a non-renewable resource. Experts don't agree on how many years Marcellus Shale drilling will occur in Pennsylvania, but many estimates are more than 20 years. In addition, other shales under Pennsylvania, such as the Utica Shale, may be economically feasible to develop, so natural gas development in Pennsylvania could be a longer process. Yet by definition, at some point the natural gas will be gone or otherwise no longer be commercially viable. The economic benefits will only last as long as the development activity occurs.

Much of the economic focus in Pennsylvania has been on the potential benefits of extracting the gas and shipping it to market; of potentially greater economic benefit may be identifying and nurturing the development in Pennsylvania of businesses that use significant amounts of natural gas in their production and thus could enjoy a comparative advantage being located close to the natural gas fields. There has been some emphasis on this in the state's economic development policy and several successes, including several natural gas power plants to be built in northern Pennsylvania, ongoing efforts to encourage bus and truck fleets to convert to CNG (the latter is funded, in part, by Act 13 Impact Fee dollars), and a proposed $1 billion plant to convert natural gas into gasoline (StateImpact 2014). One of the potentially largest value-added activities is Shell Oil's proposed "ethane cracker" plant in Beaver County. Ethane is a major feedstock for plastic production, so the hope is that other chemical-processing plants would locate near the cracker plant, bringing additional jobs to the area.

Creating such value added within Pennsylvania from natural gas has the potential of diversifying the economy away from extraction and can include significant capital expenditures which later make it difficult for firms to relocate; the Shell Oil cracker plant, for example, will cost more than $1 billion to construct (StateImpact 2014). It also could be a major source of jobs.

What Affects Economic Impacts

Development of the Marcellus Shale region affects Pennsylvania's economy through several primary means, including (1) leasing and royalty income paid to mineral right owners; (2) purchasing of services and equipment and employment by the companies directly involved in the development of the gas play (e.g., those businesses that find, extract, and process the gas); (3) employment and purchases by companies that may move to Pennsylvania because of the supply of natural gas (e.g., those businesses that want to use the gas – this is what some call "value added" from the development); and (4) negative effects of gas development on businesses, communities, and residents that affect their competitiveness and quality of life, such as loss of qualified employees to gas industry jobs, increases in local government costs, changes in environmental or water quality, health effects, and other impacts of production.

Several key elements will affect the economic impact of Marcellus, such as the timing of development, including its scale and pace. These elements are important for the full range of impacts and strongly influence the subset of impacts focused on in this study. In addition, how many of the dollars remain in the community versus immediately leave (what economists call "leakage") also plays a critical role in influencing the magnitude of the economic impacts. Each will be discussed in turn.

Timing, Scale, and Pace

The economic impacts will change throughout the development of the Marcellus Shale play, most particularly related to leasing and royalty income and workforce. Many factors will influence pace and scale, including the health of the economy as a whole, the productivity of shale wells, technological change and innovation, foreign policy, domestic energy policy, and the relative prices of different fuels. This dynamic already is visible in Pennsylvania, which has seen the pace of drilling drop from 1,961 wells in 2011 to 1,207 wells in 2013 (a 38.4 % decrease) (PA DEP 2014).

In the early years of a gas play, a large share of spending by gas companies is for lease payments to mineral right owners to acquire the right to explore and develop wells. Leasing dollars are mostly upfront, early in the development of the play as companies compete to gain control of the resource. As wells are drilled and come online, the mineral right owners receive royalty payments insofar as their wells are productive. Pennsylvania law specifies that mineral right owners must receive at least one eighth of the value of production, but some owners have negotiated for higher royalty values. The majority of these royalty dollars go to mineral right owners in the first few years of a well's active life, because production from individual Marcellus wells drops very quickly before leveling off to a slow but steady decline. This means that the majority of all the royalty dollars will be paid to mineral right owners during the active drilling phase of the Marcellus Shale play and will decline quickly once drilling ends.

Employment creation is significantly higher during the drilling phase of gas development than in the subsequent production phase, which occurs once all wells have been drilled. Brundage et al. (2010), for example, found that each wet gas well in southwest Pennsylvania requires the equivalent of 13.1 full-time employees during the year when drilling and well completion occur on the well site, spread across almost 150 occupations and 420 individuals, but only 0.18 full-time job equivalents during each of that well's subsequent producing years. Most of the employment-based economic activity from natural gas development thus will occur during the active drilling years and largely is driven by the number of wells drilled per year. This pace of drilling has important consequences for other impacts of gas development, including the need for worker housing, the number of trucks on the road, other infrastructure requirements, the quantity of water used and needing to be disposed of, and other environmental effects.

The economic impact of Marcellus Shale development within an individual community will depend upon the scale and pace of activity within that community, not necessarily the duration of drilling activity statewide. Even though some estimate that it may take 20 or more years to drill all the planned Marcellus Shale wells, the drilling phase in any one community likely will be shorter, as the crews complete work in one area before moving on to another. Whether the workers live within the communities where the drilling is occurring similarly is important, because the residence of the workers determines which municipality and school district receive their earned income tax and where the workers and their families will tend to spend much of their earnings.

Leakage

When considering the economic impacts of an activity, such as development of Marcellus Shale, it is important to track where the dollars are actually going. Money immediately leaving the community or state, such as purchases from businesses outside of the region, has less local impact than money spent at local businesses. The spatial distribution of the new dollars from Marcellus Shale activity thus can be as important as the total number of dollars involved. Leakage is particularly an issue with leasing and royalty dollars and with worker payroll. Who actually receives leasing and royalty dollars, and how those dollars are spent, has an important influence on the economic impacts of gas development. Not all mineral right owners live within Pennsylvania, so the leasing and royalty dollars they receive immediately leave the Commonwealth with little local or state impact. In addition, the Commonwealth of Pennsylvania owns and has leased a significant share of the mineral rights being developed, such as on state forest and state game land. The lease and royalty dollars from these rights go to the state government. How lease and royalty dollars are spent also has important implications for the economic impacts because it affects where, when, and how those dollars filter through the economy. Households are treating these lump-sum payments differently than regular income (Kelsey et al. 2011), with a larger proportion of such dollars being saved, invested, or spent on consumer durables than is regular income. For example, anecdotes from the drilling areas suggest many new tractors, vehicles, and four wheelers are being purchased, and many houses and barns are being repaired. In addition, how the Commonwealth spends its dollars has an economic impact. Some local governments and school districts likewise have leased their mineral rights, and their use of those dollars similarly differs from household spending.

Loss of economic impact also occurs to the extent that workers receiving wages, salaries, and other compensation spend their incomes outside of the community – an eventuality that is much more likely if they live elsewhere. Wages to transient workers typically do have some local economic impact, since such workers spend part of their income in the area where they are temporarily living (such as rent, hotel or campground fees, food, entertainment, and other basic living expenses). But since their permanent residence is elsewhere, a larger share of their earnings immediately leave the community than do wages going to local workers.

Identifying the portion of gas-related workers who are Pennsylvania residents is important from an economic impact perspective, since it affects how many wage and salary dollars remain within the Commonwealth. As with leasing and royalty dollars, from a statewide economic impact perspective, it does not matter whether workers' permanent residence is in the county where they work or if their permanent residence is elsewhere in Pennsylvania, since those dollars will circulate somewhere in Pennsylvania. Workers retaining an out-of-state permanent residence typically will spend their income differently, with a larger share immediately leaving the Commonwealth.

Direct Economic Experience

The economic impact of Marcellus Shale activity in Pennsylvania has been a source of some controversy and heated public disagreement, including how to interpret employment data (see, e.g., Herzenberg 2011; Republican Party of Pennsylvania 2011; Politics 2011; and more recently, StateImpact 2013) and the role played by overly optimistic industry-funded research (see Kinnaman 2011; Bloomburg 2012).

One contributor to such disagreements is the Center for Workforce Information and Analysis's regular publication on Marcellus Shale employment, "Marcellus Shale Fast Facts." This publication relies upon the US Bureau of Economic Analysis' Quarterly Census of Employment and Wages (QCEW) and includes a broad employment category called "Ancillary Industries" which play important roles within the shale gas development, such as general trucking, nonresidential site preparation contractors, and sewage treatment facilities. These ancillary industries are not exclusive to shale gas development, however, and instead typically are affected by multiple sectors that have nothing to do with mining or gas development. Unfortunately, it is not possible in the QCEW data to identify how many of the ancillary industry jobs relate to shale gas development, and how many relate to other economic activity. For example, the ancillary industry numbers unavoidably include drivers hauling milk between farms and milk processors, Federal Express drivers, municipal sewage treatment plant workers, and construction engineers working on residential and commercial development projects in the Philadelphia suburban area (far from the shale activity) even though none of these industries are directly affected by the drilling activity.

Despite clear caveats that these numbers include a significant number of unrelated jobs, some pro-Marcellus groups such as the Marcellus Shale Coalition, the

Republican Party of Pennsylvania (2011), and Pennsylvania Governor Tom Corbett (StateImpact Pennsylvania 2013) have included all the ancillary jobs in their "counts" of job creation associated with Marcellus Shale, while others do not include such numbers (Herzenberg 2011; StateImpact Pennsylvania 2013).

The Center for Workforce Information and Analysis numbers make clear that the mining sector (which includes natural gas development) is a very small part of Pennsylvania's overall economy (the sector's 30,031 jobs in 2014 accounted for about half a percent of Pennsylvania's 5.76 million total workforce) (Center for Workforce Information and Analysis 2014). Total employment in the sectors ancillary to shale development, which includes many jobs having no relation to Marcellus, account for about 3.7 % of the Commonwealth's total workforce, so also are a small share of overall employment statewide. From a statewide perspective, direct employment related to Marcellus Shale development is not large, even including the questionable ancillary industry numbers.

Yet Marcellus Shale has been one of the Commonwealth's major sources for new employment during the past 4 years, growing faster than most other parts of the economy. According to the Center for Workforce Information and Analysis data, while overall employment in Pennsylvania increased 3.1 % between 2009 and 2013, the core Marcellus Shale industries increased 157.4 %, and ancillary industries increased 8.2 %. The core industries accounted for 10.8 % of all the new employment in Pennsylvania (18,365 of the 170,473 net new jobs). Ancillary sector jobs accounted for an additional 9.6 % of the new jobs in the Commonwealth during this time period; not all of the growth in ancillary sectors can be attributed to Marcellus Shale activity, but such activity clearly is part of such growth.

The impact of Marcellus Shale activity in some individual counties and regions is much larger than these statewide numbers, reflecting the concentrated and very regional nature of the development activity. Indeed, many of the counties home to Marcellus Shale development are very rural with relatively small economies, and no drilling is occurring in southeast Pennsylvania, home to the largest portion of the Commonwealth's economy. So even though the employment effects may be somewhat small relative to the statewide economy, they play a much more significant role in the smaller regions where the activity is occurring. Discussing state-level economic impacts, though relevant from an overall perspective, thus can miss the magnitude of the very real impacts occurring in some locations.

The economic impacts statewide can be considered using various measures of economic activity, such as changes in wages, employment, the number of businesses, retail sales, and the flow of lease and royalty dollars to mineral right owners. Each of these will be examined in turn.

Wage Changes, 2007 to 2011

The US Bureau of Labor Statistics reports that total wages in the private sector (e.g., nongovernment jobs) decreased in Pennsylvania by 2.1 % between 2007 and 2011 after adjusting for inflation. This varied at the county level by the amount of drilling activity there; private sector wages in counties with more than 90 Marcellus Shale wells, for example, increased an average of 13.7 % during this time period, compared to a 4.9 % decline in the counties with no Marcellus wells (see Table 1). The average change at the county level in Pennsylvania during this time period was a 0.5 % decrease.

The changes in several Marcellus counties were extremely large; for example, employers in Tioga County reported a 28.9 % increase in total wages, employers in Susquehanna County reported a 30.8 % increase, and employers in Greene County reported a 38.9 % increase. These are wages paid by employers in these counties and include the wages paid to workers who live outside the county and who commute into the county.

These total wage changes translate into average weekly pay increases for workers. Counties with more wells generally experienced increases in average weekly and annual pay when adjusting for inflation. In fact, all of the 12 counties with 90 or more wells experienced positive changes in average weekly pay and average annual pay. Susquehanna County, for example, experienced an 18.7 % increase in average weekly pay and an increase of 26.2 % in average annual pay between 2007 and 2011. This is a substantially better experience than the 0.3 % increase in both average weekly and annual pay seen at the state level.

The Bureau of Economic Analysis compensation data, though measured somewhat differently than the BLS payroll information, provides a similar picture of compensation changes between 2007 and 2011 (see Table 2). Employers in counties with more Marcellus Shale activity reported employee compensation changes more positive than the state average 1.7 % decrease. Employers in counties with more than 90 wells on average reported a 13.7 % increase in total compensation, compared to a 1.9 % decrease in counties with no drilling activity. The average county experienced a 1.7 % increase during this time period.

The increases in some of the counties with Marcellus were much larger. Employers in Susquehanna County reported an increase of 25.1 % (compared to the

Table 1 Wage and income change, by drilling activity

Level of Marcellus activity in county	2007 to 2011 Change in total private sector wages, inflation adjusted (counties)	Change in average weekly pay, inflation adjusted (counties)	Change in average annual pay, inflation adjusted (counties)
More than 90 Marcellus wells	13.7 % (12)	7.7 % (12)	9.2 % (12)
10 to 89 Marcellus wells	0.1 % (11)	3.9 % (11)	4.7 % (11)
1 to 9 Marcellus wells	–3.9 % (13)	0.2 % (13)	–0.7 % (13)
No Marcellus wells	–4.9 % (31)	–0.3 % (31)	–0.8 % (31)
State average	–2.1 %	1.7 %	0.3 %
State average at the county level	–0.5 % (67)	1.9 % (67)	1.9 % (67)

Note: Total private sector wages and average annual pay information not released for Forest County due to disclosure rule
Sources: PA DEP; US Bureau of Labor Statistics, QCEW

Table 2 Worker compensation change by county 2007–2011, adjusted for inflation

Level of Marcellus activity in county	Change in total compensation (counties)	Changes by sector			
		Mining (%)	Construction (%)	Retail (%)	Transportation (%)
More than 90 Marcellus wells	13.7 % (12)	490	61.7	4.0	40.7
10 to 89 Marcellus wells	2.5 % (11)	150	25.7	–5.6	4.3
1 to 9 Marcellus wells	−1.3 % (13)	54	−1.7	–5.5	10.1
No Marcellus wells	−1.9 % (31)	16	−14.2	–6.7	5.1
State average	−1.65 %	83	−10.3	–6.8	–0.7
State average at the county level	1.7 % (67)	106	8.5	–4.3	11.4

Note: Information for some counties not released due to disclosure rule
Sources: PA DEP; US Bureau of Economic Analysis, CA06N Compensation

BLS report of 30.8 %), while in Greene County there was an increase of 33.3 % (compared to the BLS report of 38.9 %). The BEA data show particularly large increases in some sectors, especially mining, construction, retail trade, and transportation. Total compensation across Pennsylvania in the mining sector increased an average of 83.2 % at the county level when adjusting for inflation between 2007 and 2011 (see Table 2). Not unexpectedly, the counties with the most shale activity experienced the largest percentage increases in compensation in the mining sector (490 % during this time frame), while those with no Marcellus activity averaged the smallest increases (16 %). Notably, compensation in the construction sector averaged a 61.7 % increase in the highest drilling activity counties, while statewide such compensation dropped by an average of 10.3 % at the county level. A similar pattern also occurs within the retail and transportation sectors during this time frame, although the average increases are not as high as seen in the mining sector.

Pennsylvania State Personal Income Tax data shows changes in wage, salary, and other compensation changes for county residents. From a county resident perspective, the county-level numbers are less impressive. The BLS and BEA numbers are what local employers paid to workers in the county, including those working and commuting there from elsewhere, whereas the data provided by the PA Department of Revenue solely reflects earnings by residents of those counties. The Department of Revenue data show that Pennsylvania saw a 0.6 % decline in the number of residents reporting wage, salary, or other compensation income (reported as gross compensation on tax returns) between 2007 and 2011 (see Table 3), reflecting the national recession and rising unemployment. Counties with more than 90 wells on averaged experienced a similar decrease in residents reporting gross compensation income, with an average 1.1 % decline.

The state Personal Income Tax data shows total gross compensation income received by residents decreased by about 2.3 % during this 2007 to 2011 time frame. Counties with Marcellus Shale activity generally did better than the state average, with an average increase of 1.4 %. Of the 12 top counties with Marcellus activity,

Table 3 Statewide versus county-level wage and income change, by drilling activity

	2007 to 2011		
	US Bureau of Labor Statistics	PA Department of Revenue	
Level of Marcellus activity in county	Change in total private sector wages, adjusted for inflation (counties)	Change in number of returns from county residents with compensation income (counties)	Change in total compensation to county residents, adjusted for inflation (counties)
More than 90 Marcellus wells	13.7 % (12)	−1.1 % (12)	1.4 % (12)
10 to 89 Marcellus wells	0.1 % (11)	−1.3 % (11)	−0.8 % (11)
1 to 9 Marcellus wells	−3.9 % (13)	−3.4 % (13)	−4.5 % (13)
No Marcellus wells	−4.9 % (31)	−0.8 % (31)	−2.9 % (31)
State average	−2.1 %	−0.64 %	−2.3 %
State average at the county level	−0.5 % (67)	−1.4 %(67)	−2.1 % (67)

Note: Total private sector wage information not released for Forest County due to disclosure rule
Sources: PA DEP; US Bureau of Labor Statistics, QCEW; PA Dept. of Revenue, Personal Income Tax Statistics

just three saw changes below the state average. Residents in the top five drilling counties reported increases in total compensation income, ranging from 1.8 % in Lycoming County to 15.1 % in Tioga County.

When comparing the BLS data to the state income tax data in Table 3, it is important to keep in mind that the BLS data reports payroll paid by employers located in the county, irrespective of where their workers live, while the Department of Revenue data reports county residents' wage and compensation income, irrespective of where those residents work. Because many Pennsylvania residents commute to jobs outside of their home county, and local businesses' employees typically include such commuters from other counties (and residents of other states temporarily living in the county), the sets of numbers do not always align.

Employment Changes

Another way to consider the local economic impact of Marcellus Shale development is through changes in the number of jobs within the community, as observed by official federal and state data sets. Employment changes reflect the direct impact of industry spending and hiring within the community, and the additional indirect and induced employment generated in local businesses due to the industry, worker, and mineral right owner spending.

The employment data collected by the different agencies provides a generally consistent picture of job changes across Pennsylvania counties with Marcellus Shale drilling between 2007 and 2011 (see Table 4). The specific numbers reported by the

Table 4 Average employment change

Level of Marcellus activity in county	Employment change, 2007 to 2011 (includes commuters)			
	BEA (%)	BLS (%)	County business patterns (%)	Department of Revenue (County residents reporting gross compensation income on state tax returns) (%)
More than 90 Marcellus wells	3.1 (12)	3.9	2.3	1.4
10 to 89 Marcellus wells	−1.4 (11)	−5.2	−10.6	−0.8
1 to 9 Marcellus wells	−5.2 (13)	−3.2	−1.9	−4.0
No Marcellus wells	−9.2 (31)	−4.2	−4.7	−3.0
State average	−0.7	−2.3	−2.7	−2.3
State average at the county level	−4.9 (67)	−2.7	−4.1	−1.4

Sources: US Bureau of Economic Analysis CA25N; US Bureau of Labor Statistics QCEW; US Census Bureau County Business Patterns; PA Department of Revenue; PA DEP

different data series vary, in large part due to differences in data collection and the specific definitions they use, but in general they do not vary by significant amounts.

Bureau of Economic Analysis (BEA) data indicates that there was a 0.7 % average decrease in overall employment in Pennsylvania counties between 2007 and 2011. Counties with much Marcellus activity generally did better than the statewide average, and once again the top Marcellus counties on average outperformed the state. Counties with between 10 and 89 wells on average lost employment; however, they still performed better than the average Pennsylvania county during this time period.

The Bureau of Labor Statistics (BLS) data paints a similar picture of jobs and Marcellus. It indicates that Pennsylvania lost about 2.3 % of total jobs between 2007 and 2011. As in the BEA data, the top counties fared better than the statewide average, but counties with a medium level of Marcellus development performed worse than did counties with no such activity.

County Business Patterns data suggest that there was a 2.7 % average decline in jobs in Pennsylvania counties between 2007 and 2011. The counties with the most Marcellus Shale drilling generally did better than this state average, such as Tioga (9.2 % increase) and Greene (17.5 % increase). But as with the BLS data, those counties with a medium level of activity performed worse than the state average. The combination of these different economic data sources suggests that the job creation and loss experience of counties with much Marcellus Shale drilling activity between 2007 and 2011 was better for the most part, but yet not extraordinarily different than what was occurring statewide.

US Bureau of Economic Analysis data shows a similar statewide decline, with an average employment decrease of 4.9 % at the county level across Pennsylvania between 2007 and 2011 (Table 5). Employment in some of the sectors most closely related to Marcellus Shale activity performed much better, on average, than did the overall statewide economy. Mining employment increased statewide, particularly in

Table 5 Change in employment by sector, 2007–2011

Level of Marcellus activity in county	Change in total employment (counties)	Changes by sector Mining (%)	Construction (%)	Retail (%)	Transportation (%)
More than 90 Marcellus wells	3.1 % (12)	203.5	0.4	−2.2	12.0
10 to 89 Marcellus wells	−1.4 % (11)	417.7	−3.7	−10.2	−5.0
1 to 9 Marcellus wells	−5.2 % (13)	58.9	−11.8	−7.3	3.6
No Marcellus wells	−9.2 % (31)	51.2	−15.6	−5.2	−0.2
State average	−0.7	62.7	−13.6	−4.9	−1.4
State average at the county level	−4.9 (67)	158.8	−10.1	−5.9	1.9

Note: Information for some counties not released due to disclosure rule
Sources: PA DEP; Bureau of Economic Analysis, CA25N Employment

Table 6 Average flow of earnings by level of drilling activity, 2007–2011

Level of Marcellus activity	Inflow of earnings, $ in thousands, adjusted for inflation			Outflow of earnings, $ in thousands, adjusted for inflation		
	2007	2011	% change	2007	2011	% change
90 or more Marcellus wells	$ 962,843	$ 991,221	1.8	$ 442,820	$ 498,499	19.7
10 to 89 Marcellus wells	$ 161,919	$ 175,922	5.7	$ 147,673	$ 157,603	4.0
9 or fewer Marcellus wells	$ 665,334	$ 692,370	1.6	$ 1,063,806	$ 1,074,430	0.4
No Marcellus wells	$ 2,401,714	$ 2,368,135	−1.4	$ 2,258,952	$ 2,223,966	−0.8

Data sources: US Bureau of Economic Analysis CA91 (Gross Flow of Earnings); PA DEP

those counties with moderate to heavy drilling activity. Employment in the construction, retail, and transportation sectors on average similarly performed much better in high-drilling counties than the statewide experience, though retail employment on average declined more in low to moderate drilling counties than statewide or in the highest activity counties.

While Marcellus activity may be producing jobs in some of these counties, the data in Tables 4 and 5 does not address whether these positions are being filled by county residents or by those living outside the county and commuting in for work. This distinction is important as those wages being earned by a growing labor force will most likely be spent and circulated within the community the worker resides. The Bureau of Economic Analysis records the flow of income into and out of counties, and this data shows that between 2007 and 2011 counties with the most Marcellus drilling on average saw a larger increase in outflow of earnings (wages earned by out-of-county residents) than of inflow of earnings (wages earned by residents) (see Table 6). Bradford County, for example, saw a 14.5 % increase in inflow of earnings during this time but a 30 % increase in outflow of earnings. Likewise Tioga County saw an increase of 19.9 % in inflow of earnings and a 48.1 % increase in outflow of earnings. Of the 12 top counties, all but three depict this same pattern of much greater increases in outflow of earnings than of inflow of earnings.

What this suggests is that many more nonresidents than residents of Marcellus counties are taking the new jobs created by the Marcellus activity. These nonresidents may be spending some of their earnings in the county in which they work, but it is very likely that the majority of their spending is occurring in the county in which they reside. Therefore, these Marcellus county economies are not benefitting as much as they could if only residents were making up the increased labor force. In contrast, those counties without Marcellus Shale have seen more equal changes in inflow and outflow of earnings between 2007 and 2011.

Number of Businesses, 2007 to 2011

Another way to consider the economic impact of Marcellus Shale is how the number of local businesses is changing during development of the shale. County Business Patterns data indicate that Pennsylvania lost about 3.3 % of its private sector businesses between 2007 and 2011, but this varied between individual sectors. The number of mining sector businesses (which include natural gas development companies) increased by 25 % statewide during this time period, while the number of construction businesses fell by 11.5 %, and retail businesses fell by 4.6 % (see Table 7).

Marcellus Shale counties generally did somewhat better than the state as a whole in retaining and adding businesses. Counties with the highest levels of Marcellus Shale activity on average performed much better than the statewide average, such as a 128.8 % increase in mining-related businesses and a 23.9 % increase in transportation-related businesses. Changes in the number of construction, retail, and

Table 7 Average change in number of businesses

Level of Marcellus activity in county	Change in number of businesses, 2007 to 2011 (percent change)					
	County business patterns					BLS
	All sectors (%)	Mining/ gas (%)	Construction (%)	Retail (%)	Transportation (%)	All sectors (%)
More than 90 Marcellus wells	−0.7	128.8	−7.5	−3.2	23.9	3.7
10 to 89 Marcellus wells	−4.3	32.9	−8.2	−4.9	−1.5	−0.3
1 to 9 Marcellus wells	−4.9	12.9	−7.5	−8.1	−5.1	0.7
No Marcellus wells	−4.4	−0.5	−13.1	−5.2	−1.9	−0.6
State average	−3.3	25.0	−11.5	−4.6	1.0	2.2
State average at the county level	−3.8	31.7	−10.2	−5.4	2.2	0.5

Sources: PA DEP; US Census, County Business Patterns; Bureau of Labor Statistics, QCEW

transportation firms varied across the Marcellus counties, with some doing better and others worse than the state.

A similar data series on the number of businesses, as reported by the Bureau of Labor Statistics, appears a bit more positive than County Business Patterns, showing an average 2.2 % increase in total businesses at the county level statewide between 2007 and 2011 (Table 7). This difference occurred, in part, due to how the two series collect their data; County Business Patterns identified 295,720 businesses statewide in 2011, while the BLS identified 331,723 businesses. The general trends in the BLS data are similar to the County Business Patterns data; counties with much Marcellus activity generally did a bit better than the state average.

Retail Sales

As residents gain lease and royalty income and new jobs are created from Marcellus Shale development, more money should be flowing into local retail stores. State sales tax collection data provide a perspective on how the local retail sector has been changing during Marcellus development. Though not a perfect correlation to retail sales activity because the tax excludes food and clothing purchases, collections rise and fall with the level of retail activity.

The Pennsylvania Department of Revenue data indicates that sales tax collections decreased 4.8 % statewide between July 1, 2007, and June 30, 2013, but counties with Marcellus activity on average did better; counties with the most shale development on average experienced a 14.2 % increase in sales tax collections. Meanwhile, collections in counties with no Marcellus activity during this time period fell an average of 13.1 % (see Table 8). The changes in several counties were particularly dramatic; collections in Bradford County increased 45.4 % and in Greene County increased 73.6 % during this time frame (adjusted for inflation). These findings are consistent with those of a survey of businesses in two Marcellus Shale counties that found 35 % of the businesses in Bradford County and 25 % of the businesses in Washington County reported sales in increases due to Marcellus Shale activity (Ward and Kelsey 2011). The Department of Revenue data suggests

Table 8 Average change in state sales tax collections, by Marcellus activity

Level of Marcellus activity in county	Percent change (inflation adjusted), July 1, 2007, to June 30, 2013 (number of counties at level of Marcellus activity)
150 or more Marcellus wells	14.2 % (10)
10 to 149 Marcellus wells	–5.2 % (17)
1 to 9 Marcellus wells	–5.7 % (12)
No Marcellus wells	–13.1 % (28)
State average	–4.8 %
State average at the county level	–5.7 % (67)

Sources: PA DEP; PA Department of Revenue

that Marcellus Shale development is positively affecting the local economies in Marcellus counties by increasing spending in the retail sector.

Leasing and Royalty Dollars, 2007 to 2010

Who receives leasing and royalty dollars and how those dollars are spent has an important influence on how much of the economic impact of Marcellus Shale development goes to the local community. In Pennsylvania, as in most other states, surface land owners do not necessarily own the mineral rights under their land. Surface and mineral rights can be severed and be owned (and sold) separately from each other. This is relatively common in areas of the Commonwealth which historically have experienced coal mining and natural gas or petroleum development, such as western Pennsylvania counties. Because this mineral owner information typically is not formally aggregated or tracked, other than on a deed-by-deed basis, no one really knows where the leasing and royalty dollars are going. Kelsey et al. (2011) used GIS analysis of land ownership patterns and found that approximately 51 % of the land with Marcellus Shale in Marcellus counties is owned by residents within the county. If the distribution of mineral rights and land ownership is similar, this means about half of the total leasing and royalty dollars being generated within a county go to residents of that county. The other 49 % of total leasing and royalty dollars immediately leave the county, with 25 % going to owners living elsewhere in Pennsylvania, 8 % to owners living outside Pennsylvania, and the remaining 17 % going to the public sector, primarily the Commonwealth.

Yet when discussing this finding, Kelsey et al. noted that this 51 % estimate very likely overestimates the actual amount of lease and royalty dollars going to county residents. Many of the mineral rights in counties with past coal or gas development were severed generations ago during that earlier activity and have subsequently been passed down through families, splintering into multiple ownerships across children and grandchildren. Given the relatively high amount of outmigration from Pennsylvania over the past decades, it is likely that many of the current mineral right owners do not live in the Commonwealth. Companies also purchased mineral rights during that time period, so the rights no longer are owned by individuals.

Of the lease and royalty dollars going to county residents, not all are spent, or are spent locally, which reduces their local economic impact. Kelsey et al. (2011) found that mineral owners are saving about 55 % of the total leasing dollars and 66 % of the royalty dollars they receive and that they are spending these dollars differently than their regular income; the respondents indicated that much of their spending is on motor vehicles, state and federal taxes, and real estate. Follow-up economic impact analyses of these consumer spending patterns in Bradford, Sullivan, Susquehanna, Tioga, and Wyoming counties found that much of this spending is being done outside of the counties where the mineral owners live, reducing the local economic impact (Kelsey et al. 2012a, b, c, d, e). Small rural counties like these have fewer

Table 9 Percent change in rents, royalties, patents, and copyrights income by drilling activity

Level of Marcellus activity in county	2007 to 2011, in $ thousands Average change in rents, royalties, patents, and copyrights, adjusted for inflation (percentage increase)	Average change in number of returns reporting this income (%)
More than 90 Marcellus wells	$60,298 (189.9 %)	35.3
10 to 89 Marcellus wells	$11,220 (74.2 %)	8.73
1 to 9 Marcellus wells	$13,613 (28.1 %)	18.16
No Marcellus wells	$4,391 (6.3 %)	10.33
State average at the county level	$17,315 (34.7 %)	15.71

Sources: PA DEP; PA Department of Revenue, Personal Income Tax Statistics

local shopping alternatives, which means their residents are more likely to shop outside of the county.

State tax information from the Pennsylvania Department of Revenue provides some concrete numbers about how resident income is changing in the Marcellus counties and on spending in local businesses. The number of state personal income tax forms reporting rents, royalties, patents, and copyrights income (how lease and royalty dollars from Marcellus Shale are categorized) increased by an average of 15.7 % at the county level statewide between 2007 and 2011 (see Table 9), which means that there was an increase in the number of Pennsylvanians receiving such income. The percent changes in the number of returns reporting such income were much higher in counties with significant Marcellus drilling activity, including an 88.8 % increase in Bradford County, a 78.2 % increase in Tioga County, and a 62 % increase in Washington County.

In the time frame of 2007 to 2011, rents, royalties, patents, and copyrights income received by residents increased by an average of 34.7 % at the county level statewide (Table 9). The dollar and percentage increases were very large, particularly in the counties relatively new to gas development. Taxpayers in Bradford County, for example, reported an increase of $138.7 million (953.2 %) in such income, while taxpayers in Tioga County reported an increase of about $80.1 million (857.1 %) in such income between 2007 and 2011.

Residents of counties which have had a longer history of natural gas and coal development typically reported smaller increases in total lease and royalty income. In counties with 90 or more Marcellus wells and little past history of gas development, $213,273 in rents, royalties, patents, and copyrights income was received per Marcellus well, compared to $198,459 per Marcellus well in counties with 90 or more Marcellus wells and past history of gas development (see Table 10). Residents of counties with the most drilling without history of coal or gas extraction on average experienced an increase of 641.6 % in income of this type, while residents of those high-activity counties with history of energy extraction experienced an increase of 156.9 %. This same pattern is true of counties with 10–89 Marcellus wells, although there is just one county (Sullivan County) which has this level of Marcellus activity and has no past history of coal or natural gas extraction.

Table 10 Change in rents, royalties, patents, and copyrights income reported by residents, by past history of coal and gas extraction in the county, 2007 to 2011 income adjusted for inflation

	Counties by past history of coal and gas development $ thousands (counties in classification)	
	Little or no past history *much less likely to have surface and mineral rights to be owned by different owners*	**Past history** *more likely to have surface and mineral rights to be owned by different owners*
Counties with 90 or more Marcellus wells		
Average change in income per well	$213,273 (5)	$198,459 (7)
Average change in income reported	641.6 %	156.9 %
Counties with 10 to 89 Marcellus wells		
Average change in income per well	$604,982 (1)	$305,691 (10)
Average change in income reported	1886.3 %	106.6 %
Counties with 1 to 9 Marcellus wells		
Average change in income per well	$3,396,230 (8)	$3,604,216 (5)
Average change in income reported	35.8 %	39.9 %

Data sources: Pennsylvania Department of Revenue; PA DEP

Changes in rents, royalties, patents, and copyrights income are relatively similar across counties with 1–9 Marcellus wells, regardless of their history of coal and natural gas extraction.

Multiplier Effects on Employment and Income

The raw employment, wage, and tax numbers related to drilling activity do not wholly represent the broader economic impacts of shale gas development, such as on other businesses providing services and supplies to the industry, spending by workers, and other associated impacts. Such indirect and induced impacts of natural gas development have commonly been studied using the economic input–output model IMPLAN (see, e.g., Center for Business and Economic Research 2008; Considine et al. 2009, 2010; Kelsey et al. 2011; National Energy Technology Lab 2010; Pennsylvania Economy League 2008; Scott and Associates 2009). Yet there are clear cautions to its use and interpretation for natural gas development, including carefully taking into account the flow of royalty dollars (Kay 2011; Kinnaman 2011).

In their study of the economic impacts of Marcellus Shale in Pennsylvania, Kelsey et al. (2011) used Geographic Information System (GIS) data, a survey of landowners to identify where and how such dollars are spent, and sensitivity analysis about nonresident employees to address some of these identified limitations of input–output analysis. Their study was funded by the Pennsylvania Department of Community and Economic Development in response to the industry-funded highly optimistic and controversial projections of Considine et al. (2010) and was intended as an independent assessment of the potential economic impacts of Marcellus Shale activity in Pennsylvania.

Kelsey et al. found that about 7.7 % of the land area[1] in Marcellus Shale counties of Pennsylvania is owned out of state, which means that such lease and royalty dollars would immediately leave Pennsylvania without creating additional economic impacts in the Commonwealth. An additional 17 % of the land area is owned by the public sector, primarily the state government, so the lease and royalty dollars associated with this land flow to state government rather than to individual landowners.

To take into account how lease and royalty dollars are actually spent, they surveyed 1,000 randomly selected landowners living within 1,000 feet of an active Marcellus Shale well and had a 50.1 % response rate. Importantly, as referenced earlier, Kelsey et al. (2011) found landowners were saving about 55 % of the leasing dollars, and 66 % of the royalty dollars they had received. The dollars actually spent by respondents were atypical of consumer spending, which suggests the owners view these funds as a windfall.

The role of nonresident workers filling the local jobs has been a sensitive subject in many of the communities with shale gas development. In addition to affecting employment opportunities for local residents, such workers take much of their money to their home communities, reducing the economic impact within the shale region. Kelsey et al. relied upon the results from a 2010 workforce needs assessment (Brundage et al. 2010), which found that about 37 % of the Marcellus workforce were non-Pennsylvania residents, to account for such leakage.

Using this data, Kelsey et al. estimated the total economic impact of Marcellus Shale development activity in Pennsylvania in 2009 included 6,741 direct jobs resulting from industry spending and an additional 2,631 indirect jobs, for a total of 9,372 jobs directly related to industry spending. The overall economic impact, which includes these jobs plus the effects of lease and royalty dollars, thus ranged between 23,385 and 23,884 jobs and $3.1 and $3.2 billion (see Table 11). These findings are consistent with several other Marcellus Shale employment studies which relied upon different methodologies, including company interviews about employment needs (Brundage et al. 2011), direct observation of hiring and employment trends

Table 11 Summary of economic impacts and total economic impact, 2009

Impact type	Employment	Labor income	Value added	Output
Lower bound: if 50 % of nonresident employee income stays in PA and 15.4 % of mineral rights are owned out of state				
Total economic impact	23,385	$1,202,855,556	$1,863,290,275	$3,138,994,978
Upper bound: if 75 % of nonresident employee income stays in PA and 7.7 % of mineral rights are owned out of state				
Total economic impact	23,884	$1,225,210,536	$1,897,448,298	$3,195,740,526

Source: Kelsey et al. 2011

[1]They noted that because public data on mineral right ownership is unavailable, they had to use surface ownership as a proxy for mineral right ownership. Due to the past history of gas and oil development in some Pennsylvania counties, many of these rights were severed generations ago, however. They argue that due to population outmigration over the decades, using surface ownership as a proxy likely overestimates the actual proportion of mineral rights owned by Pennsylvanians.

Table 12 Estimated total statewide employment impacts related to Marcellus Shale activity

Year	Marcellus wells drilled	Total employment
2009	817	23,884
2010	1,602	46,833[a]
2011	1,961	57,327[a]
2012	1,350	39,466[a]
2013	1,207	35,285[a]

Data sources: PA DEP
[a]Estimated using Kelsey et al. 2011, per well calculations

(Herzenberg 2011), and comparisons of US Bureau of Economic Analysis data from high-drilling and non-drilling counties (Weinstein and Partridge, 2011). Brundage et al. estimated that 8,752 direct jobs were created as a result of industry spending on drilling activity in Pennsylvania during 2009. Herzenberg calculated there were 9,288 new jobs within the Marcellus Shale industry between 2007 and 2010, while Weinstein and Partridge estimated a net gain of 10,000 direct and indirect jobs in the natural gas industry between 2004 and 2010.

The Kelsey et al. economic impact study estimated that total employment per well drilled is around 29 full-time equivalents. This includes the direct and indirect impacts of industry spending, plus the employment impacts of new household spending resulting from higher employment, and the lease and royalty income. Brundage et al. (2011) similarly estimated per well employment but focused solely on the direct impacts, so understandably is smaller at 13.1 full-time equivalents per well.

The full employment impact per well estimated by Kelsey et al. (2011) would suggest that total employment statewide related to Marcellus Shale development in 2013 was around 35,285 workers (see Table 12), down from a peak employment of 57,327 in 2011 when Marcellus Shale drilling activity was at its highest in the Commonwealth. These estimates should be treated very cautiously because of countervailing trends affecting the statewide economic impact; increasing industry productivity since 2009 likely has dropped employment needs per well, reducing the per well economic impact. At the same time, the increasing emphasis on training and hiring Pennsylvania workers since 2009 should have increased the amount of employee wages remaining within Pennsylvania, so would have increased the statewide per well economic impact. Without additional analysis, it isn't possible to determine which trend has had a larger impact and thus whether total employment impacts per well have increased or decreased.

Discussion/Implications

What the Numbers Show

The state and federal economic data series depict a relatively consistent perspective on how Marcellus Shale activity is affecting local economies in Pennsylvania. The official data suggests that Marcellus Shale is having generally positive effects on employment, wages, and local business activity. Much of the economic impact, even in the counties with the most drilling, however, appears relatively modest compared to the dramatic scale claimed by some industry-funded economic impact studies, such as the 139,889 jobs statewide in 2010 (Considine et al. 2011) and in some of the political rhetoric over shale gas development.

Wages and Income Are Up Much More than Employment

The data show that wages and income (Tables 1 and 2) generally increased more than did the number of workers (Tables 3). Between 2007 and 2011 in Bradford County, for example, total wages and salaries increased 23.9 %, while total employment increased by 8.4 % (based on BEA data sets). This suggests that much of the impact on workers has been more work hours, higher pay, or a combination of both, rather than significant new job creation. This result is consistent with many of the anecdotes from counties with drilling and should not be too surprising given the relatively small size of many of these counties; there are not that many qualified unemployed workers in these counties compared to the potential labor needs.

Increases in Local Employment

The BEA and BLS data (Table 4) show that the number of workers in many counties with Marcellus has been increasing despite a statewide average decrease in employment between 0.7 (BEA data set) and 2.3 % (BLS data set). As the Center for Workforce Information and Analysis notes, Marcellus counties generally have had lower unemployment rates than other Pennsylvania counties, which is supported by both the BEA and BLS findings.

More importantly, the BEA and BLS data report employment in the county, which includes workers who commute into the county from elsewhere. How many of these jobs are going to county residents and how many are going to nonresidents is not clear from the numbers. The personal income tax data collected by the Pennsylvania Department of Revenue from county residents suggests that there has been only a modest increase in the number of county residents receiving wages or salaries (Table 4) despite the number of new jobs being created. It is clear from comparing these data sets that a large portion of the increase in employment seen in the Marcellus counties is actually a reflection of increased workers that commute into these counties to work but reside elsewhere.

Large Increases in Some Sectors

The data indicates that some sectors in the county economies are doing particularly well with Marcellus Shale activity in regard to employment, particularly the mining/oil, construction, retail, and transportation sectors (Table 2). There have been substantial increases in employment within the mining sector, as well as the transportation sector, and likewise large increases in worker compensation within these industries (Table 4).

The data reflecting impacts within the retail sector of county economies is different. It appears that while wages to workers has increased (Table 2) and spending has increased (Table 8), actual employment within the retail sector on average decreased in the low and moderate drilling counties more than statewide (Table 5). These data suggest that the impact on local retail businesses has been an increase in spending which has prompted an increase in wages to workers and not an increase in retail employment.

Leakage

That employment increases in counties with Marcellus Shale activity generally are much more modest than some would expect, given the large spending by industry, implies that much of the economic benefits of shale gas development are leaking from the communities and state. There is no doubt that there is much physical activity occurring in Pennsylvania counties due to Marcellus, but this does not guarantee strong connections with the local or state economy, any more than do high levels of traffic on an interstate highway guarantee economic activity in a community bisected by that highway.

When many of the workers and companies are based outside the counties, and much of the equipment and supplies they use are brought in from elsewhere, the connections to the local economy are not as strong as they would be otherwise. The amount of outside workers, businesses, and suppliers can be viewed as significant opportunities for increasing the local economic development impacts of Marcellus Shale, insofar as local workers and firms can "capture" a larger proportion of such activity. The data indicate that so far local efforts have not reached their full potential.

Value Added

As discussed earlier, much of the potential employment and economic impacts of shale gas development likely is within the businesses that will use the natural gas as an input for production. The Commonwealth has been exploring and promoting such value-added businesses, such as fleet use of CNG, but there is little doubt that more could be done to encourage the growth of such businesses with Pennsylvania. In addition to creating added economic benefit from the drilling, such development can help diversify the economy, evening out some of the instability and risk associated with drilling and extraction.

Need to Think Long Run

Shale gas is a nonrenewable resource, which means that it and its associated economic impacts will be depleted at some point in the future. In addition, prior studies indicated that the majority of employment impacts will occur during the early drilling phase of gas development (Brundage et al. 2011), not during the longer-term gas production itself. Royalty income to landowners similarly will be highest during the early drilling phase because production per well drops rapidly. The recent major drop in drilling activity in Pennsylvania demonstrates that price volatility by itself can create wide short-term fluctuations in activity and thus employment and income.

The Commonwealth and its communities thus need to view Marcellus Shale development as a temporary and shifting boost to the economy, rather than as a long-term economic development strategy. Once the gas drilling ends, most of the direct economic impacts similarly will end. This means the Commonwealth, communities, and residents need to work actively to tailor development so it leaves the economy better off in the long term, such as maintaining the local quality of life, making sure current infrastructure investments have long-term usefulness (and are paid off before the boom slows), encouraging the creation of local businesses that broaden the economy so it is less dependent in the long run upon gas development, minimizing the impact on non-gas parts of the economy, and protecting the water, air, and forest ecosystems that future generations will depend upon.

The long-run economic impacts of Marcellus Shale development, particularly for resource-dependent sectors of the economy like tourism and agriculture, likely will be very different than what occurs in the early years of development due to cumulative and scale effects as the number of wells drilled and in operation increase. Some have argued that tourism will decline (either because of actual physical changes to the landscape or because controversy over drilling scares tourists away), though others have argued that tourism may increase because access roads and pipeline rights of way are opening up previously inaccessible hunting lands and creating better ecosystems for white-tailed deer, which could attract more hunters.

In addition, most of the existing uncertainty about Marcellus Shale development relates to its possible long-term effects, including water quality, land use, forest, health, and social impacts. There is uncertainty about how similarly the economic activity will conform to the boom/bust cycles that have occurred with energy development in the west and which have characterized Pennsylvania's prior experience with timber, coal, and petroleum development. Much of this depends upon the scale and pace of the development, plus whether there are unforeseen cumulative effects as the play is developed and the number of wells (and supporting access roads, miles of pipeline, and other infrastructure) increases. In addition, it depends upon how individuals and communities respond (e.g., to what extent will recipients of leases and royalties sell the surface rights and move away with that stream of income, taking the economic benefit with them? Will communities use the current economic benefits to strategically invest for the future?) and whether the gas is mostly exported and used out of state, or if it instead is used to attract other industries and thus helps build a more diversified and strong economy in the Commonwealth. No one knows the answers to these questions because much of this will occur in the future, but it is important to be gathering appropriate information now so we can predict and anticipate these earlier rather than later. In addition, local, state, and federal policy will influence this future.

Unfortunately, planning for the long run can be inconsistent with the short-term election cycles of local, state, and national policymakers. The long run can require foregoing short-term gains or problem-solving. Recent Pennsylvania politics suggest that the short term often wins out; both a democratic and a republican governor have used leasing state forest land as a way of balancing the state budget (or beefing up the General Fund). It can take real political leadership to forego short-term temporary gains for long-run benefits. North Dakota's Legacy Fund is a clear example of what can be done with enough political will and leadership; the Fund was created in 2010 through a voter approved constitutional amendment and is composed of 30 % of oil and gas production and extraction taxes collected by the state government. As of April, 2014, the fund has accrued about $2 billion and is growing by $700 million a year (Slate 2014). The fund was created in a manner to ensure that the monies cannot be squandered quickly. The money cannot be touched until 2017, and afterwards the principal can only be spent by a two-thirds majority of the North Dakota House and Senate (with at most 15 % of the principal spent in any 2-year period).

So far, Pennsylvania politicians have not shown similar courage to set aside funds for the future. Act 13 of 2012 created a Legacy Fund in name, yet little is actually put aside for future needs but rather is allocated to state agencies and local governments for immediate use. In addition, dollars legislatively set aside for long-term environmental needs through the Oil and Gas Lease Fund were taken to balance at least two state budgets (StateImpact 2011).

Distributional Issues

Some would argue that the distribution of the economic benefits and the costs of Marcellus Shale development is important to consider, particularly how the economic benefits occurring within communities with drilling compare to the local costs of such drilling. There is little doubt that development activity creates nuisances, inconveniences, and risks for residents where drilling is occurring. How much of the economic benefits remain within the communities with gas development activity and how these are distributed are important from a social justice perspective. The 2011 Kelsey et al. economic impact study suggested that at least through 2010, a large proportion of the economic benefits were leaving the communities with drilling activity and the Commonwealth.

The distribution of benefits within the communities themselves also is important from a justice perspective. The data suggests that much of the lease and royalty income is going to a small proportion of the local population. State personal income tax data from the most active Marcellus Shale counties indicates that increases in lease and royalty income exceed increases in wage and net profit income (Hardy and Kelsey 2013), and yet such dollars flow to a small share of the population. In 2010, for example, in Bradford County, 19.2 % of tax returns reported receiving such income, as did 14.3 % of returns from Susquehanna County and 10.9 % from Greene County. Ownership of land, and thus the flow of lease and royalty dollars, itself is highly concentrated, with the top 10 % of local landowners in the most active Marcellus Shale counties typically owning between 72 % and 88 % of the locally owned acreage (Kelsey et al. 2012a).

Costs

As with most prior economic studies of Marcellus Shale, this study was unable to directly consider the possible economic costs associated with the development, even though some argue that these costs may be significant. Such a comparison of local costs and of local benefits is essential to be able to determine whether the community and residents are better or worse off as a result of Marcellus Shale activity.

The total employment and income numbers in the data include all sectors within the counties, including those who may have been harmed by Marcellus Shale activity, so any possible short-run economic costs are reflected in the data. Several studies suggest that high levels of gas development can harm local economies in the long run; Headwaters Economics (2008) examined energy-dependent counties in the western United States and found that energy-dependent counties lag behind the economic performance of more diversified nearby local economies. Weinstein and Partridge (2011) note that the demand for labor during the initial boom can crowd out other sectors, and rising housing costs can displace low-income workers, leaving the local economy less diverse and more vulnerable to economic shocks. The short- and long-run costs are important to consider for an overall view of the economic implications of Marcellus Shale development.

Most importantly, this study only focused on the job and income effects of gas industry spending. These economic elements must be balanced by an understanding of the costs of such development. Existing economic impact studies of Marcellus development, including this one, have focused almost exclusively on job and income creation resulting from gas industry spending, including leasing and royalty payments, payroll, and purchases from other businesses. In contrast, no economic study so far has included the potential costs of Marcellus Shale development, such as the impact on existing businesses losing employees due to Marcellus activity, effects on human health, damage and cleanup costs resulting from accidents or environmental degradation, changing state and local government costs due to activity, and rising cost of living (such as rents) in the counties experiencing drilling.

There clearly are and will be costs associated with Marcellus Shale development, both out-of-pocket and nonmonetary (such as the ecosystem effects of forest fragmentation or water quality impacts, impacts on human health, and other effects). There may also be opportunity costs, such as businesses that choose not to locate or expand within Pennsylvania due to the changes resulting from Marcellus Shale development. Yet because Pennsylvania is still relatively early in the Marcellus play, these currently cannot be fully identified or quantified. Some costs may not show up until much later in the development of the play, such as when the amount of activity passes currently unknown thresholds or achieves a critical mass. That the costs currently cannot be comprehensively measured does not mean that such costs do not or will not exist, but rather means it is vital to investigate and identify them. To focus only on jobs, income, or tax revenue without putting those into a broader context can be very misleading and costly in the long run.

Caveats

Some caution is required when interpreting these numbers. Changes at the county and state level reflect everything that occurred within the county, not just Marcellus Shale-related activity. Shifts between sectors in the local economy may be hidden if hiring in one sector counterbalances layoffs in another. The methodology in this study is the same as used by the Pennsylvania Department of Labor and Industry's Center for Workforce Information and Analysis in its monthly "Marcellus Fast Facts" (although they focus on sector-level and regional changes, not county level). Though imperfect, comparing the county-level results to statewide averages provides some independent perspective on overall economic trends and what might have occurred in the county in the absence of Marcellus development.

The national economy entered into its recession during the development years of Marcellus Shale, which makes local employment changes more volatile and more difficult to interpret relative to Marcellus Shale activity. The state average changes provide some measure of reasonable comparison because the statewide trends include broader impacts than just the natural gas drilling activity. Weinstein and Partridge note that the gas sectors' small share of total Pennsylvania employment is "simply not enough to have a significant effect on total jobs and on unemployment in the state," which means Marcellus Shale development's impact on the statewide averages would not be very significant, so comparisons to the state average are appropriate and meaningful.

Conclusions

It is clear from federal and state employment and income data that there is a lot of local economic activity occurring in the communities with drilling activity, but the experience is uneven across counties, seemingly not directly related to the level of drilling activity. Mineral owners in these counties report significantly large amounts of new lease and royalty income, which is of clear benefit to the owners. The mining, construction, retail, and transportation sectors generally are doing well. Sales tax collections are up significantly in some counties, indicating significant increases in local retail spending. Wages and salaries paid by local employers similarly have increased significantly in some counties, particularly within the mining, construction, and transportation sectors. Counties with Marcellus activity typically did a bit better retaining or adding local businesses than did the rest of the state. Yet with a few exceptions, the number of county residents reporting wage and salary income, and the number of employees working for businesses within these counties, is not up dramatically.

Many of these economic numbers appear more modest than would be expected, given the billions of dollars being spent to develop the Marcellus Shale. The county-level economic numbers do not negate the broader economic impacts which are occurring in neighboring counties, elsewhere in Pennsylvania, and nationally as a result of Marcellus Shale development. They instead suggest that a large proportion of the employment and other economic impacts from Marcellus Shale are occurring outside the counties where the drilling is occurring, despite the relatively high level of activity in those counties. This means there are significant opportunities for increasing the local economic development impacts of Marcellus Shale but that so far local efforts have not reached their full potential. It also raises questions about how the local economic benefits of Marcellus Shale development compare to the local costs and inconveniences in the counties where drilling is occurring and where the other economic benefits are going.

How long Marcellus Shale development will last in Pennsylvania is unclear, particularly with the current uncertainty about when natural gas prices will rise, but industry planning indicates it likely will be decades. It is critical to remember that natural gas development is a nonrenewable resource, so by definition drilling will end at some point and so will its local and statewide economic impacts. The long-run implications of Marcellus Shale development are still unknown. Jobs and income in the short run are important, but many would argue that other factors are equally (if not more) important, such as clean water, healthy forests and other ecosystems, clean air, and public health. In addition to affecting quality of life, these are important resources for the future of Pennsylvania communities, including future economic opportunities, social and physical infrastructure, well-functioning local government and institutions, and community well-being.

The challenge and opportunity for residents, local businesses, and leaders in Pennsylvania is to find ways of using the current Marcellus-related activity to strengthen the Commonwealth and economy for the long run, so when the drilling and natural gas production ends, the Commonwealth and its residents are better off than they were before the gas development began.

References

Bloomburg.com (2012) Frackers Fund University Research that proves their case. Available at: http://www.bloomberg.com/news/2012-07-23/frackers-fund-university-research-that-proves-their-case.html Accessed 6 June 2014

Brown JP (2014) Production of natural gas from shale in local economies: a resource blessing or curse? Econ Rev. Forthcoming

Brundage TL, Jacquet J, Kelsey TW, Ladlee JR, Lobdell J, Lorson JF, Michael LL, Murphy TB (2011) Pennsylvania Statewide Marcellus Shale workforce needs. Marcellus Shale Education and Training Center, Williamsport

Brundage TL, Jacquet J, Kelsey TW, Ladlee JR, Lobdell J, Lorson JF, Michael LL, Murphy TB (2010) Southwest Pennsylvania Marcellus Shale workforce needs assessment. Marcellus Shale Education and Training Center, Williamsport

Center for Business and Economic Research (2008) Projecting the economic impact of the Fayetteville Shale Play for 2008–2012. Sam M. Walton College of Business, Fayetteville

Center for Workforce Information and Analysis (2014) Marcellus Shale fast facts: May 2014 edition. Pennsylvania Department of Labor and Industry, Harrisburg

Considine TJ, Watson R, Entler R, Sparks J (2009) An emerging giant: prospects and economic impacts of developing the Marcellus Shale natural gas play. The Pennsylvania State University, Department of Energy and Mineral Engineering, University Park

Considine TJ, Watson R, Blumsack S (2010) The economic impacts of the Pennsylvania Marcellus Shale natural gas play: an update. The Pennsylvania State University, Department of Energy and Mineral Engineering, University Park

Hardy K, Kelsey TW (2013) Marcellus Shale and local economic activity: what the 2012 Pennsylvania State tax data say. The Pennsylvania State University, Penn State Extension, University Park

Headwaters Economics. 2008 Energy Revenue in the Intermountain West: State and Local Government Taxes and Royalties from Oil, Natural Gas, and Coal, Energy and The West Series. http://headwaterseconomics.org/pubs/energy/HeadwatersEconomics_EnergyRevenue.pdf

Herzenberg S (2011) Drilling deeper into job claims: the actual contribution of Marcellus Shale to Pennsylvania job growth. Keystone Research Center, Harrisburg

Jacobsen GD, Parker DP (2014) The economic aftermath of resource booms: evidence from boomtowns in the American West. Econ J. Forthcoming

James A, Aadland D (2011) The curse of natural resources: an empirical investigation of U.S. Counties. Res Energy Econ 33(2):440–453

Kay DL (2011) The economic impact of Marcellus Shale gas drilling: what have we learned? What are the limitations? Cornell University, Ithaca

Kelsey TW, Shields M, Ladlee JR, Ward M (2011) Economic impacts of Marcellus Shale in Pennsylvania: employment and income in 2009. Marcellus Shale Education and Training Center, Willamsport

Kelsey TW, Metcalf A, Salcedo R (2012a) Marcellus Shale: land ownership, local voice, and the distribution of lease and royalty dollars, Center for Economic and Community Development White Paper Series. Penn State University, University Park

Kelsey TW, Shields M, Ladlee JR, Ward M (2012b) Economic impacts of Marcellus Shale in Bradford County: employment and income in 2010. Marcellus Shale Education and Training Center, Williamsport

Kelsey TW, Shields M, Ladlee JR, Ward M (2012c) Economic impacts of Marcellus Shale in Sullivan County: employment and income in 2010. Marcellus Shale Education and Training Center, Williamsport

Kelsey TW, Shields M, Ladlee JR, Ward M (2012d) Economic impacts of Marcellus Shale in Susquehanna County: employment and income in 2010. Marcellus Shale Education and Training Center, Williamsport

Kelsey TW, Shields M, Ladlee JR, Ward M (2012e) Economic impacts of Marcellus Shale in Tioga County: employment and income in 2010. Marcellus Shale Education and Training Center, Williamsport

Kelsey TW, Shields M, Ladlee JR, Ward M (2012f) Economic impacts of Marcellus Shale in Wyoming County: employment and income in 2010. Marcellus Shale Education and Training Center, Williamsport

Kinnaman TC (2011) The economic impact of Shale gas extraction: a review of existing studies. Ecol Econ 70:1243–1249

Marcellus Shale Coalition (2011) The Marcellus multiplier: powering Pennsylvania's supply chain Marcellus Shale Coalition, Cannonsburg

National Energy Technology Lab (2010) Projecting the economic impact of Marcellus Shale gas development in West Virginia: a preliminary analysis using publicly available data. U.S. Department of Energy, Morgantown

Papyrakis E, Gerlagh R (2007) Resource abundance and economic growth in the United State. Eur Econ Rev 51(4):1011–1039

Pennsylvania Department of Environmental Protection (2011) 2010 Wells Drilled By County as of 02/11/2011. Pennsylvania Department of Environmental Protection, Harrisburg

Pennsylvania Department of Revenue (2013) Tax compendium 2007–08; through 2011–2012. Pennsylvania Department of Revenue, Harrisburg

Pennsylvania Department of Revenue (2014) Personal income statistics, 2007 through 2011. Pennsylvania Department of Revenue, Harrisburg

Pennsylvania Economy League (2008) The economic impact of the oil and gas industry in Pennsylvania. Pennsylvania Economy League, Pittsburgh

Politics PA (2011) PR battle continues over Marcellus jobs numbers. http://www.politicspa.com/pr-battle-continues-over-marcellus-jobs-numbers/25734/. Accessed 4 June 2014

Republican Party of Pennsylvania (2011) PA GOP: liberal 'Think Tank' Keystone Research Center Releases Political Attack on Job Creation. http://www.pagop.org/2011/06/pa-gop-liberal-%E2%80%9Cthink%E2%80%9D-tank-keystone-research-center-releases-political-attack-on-job-creation/. Accessed 5 June 2014

Scott, LC and Associates (2009) The economic impact of the Haynesville Shale on the Louisiana economy in 2008. Louisiana Department of Natural Resources, Baton Rouge

Slate (2014) Little Sovereign Wealth Fund on the Prairie. May 29. http://www.slate.com/articles/business/the_juice/2014/05/north_dakota_sovereign_wealth_fund_the_state_holds_tight_to_its_sense_of.html. Accessed 5 June 2014

StateImpact Pennsylvania (2011) Can Pennsylvania's State Forests Survive Additional Marcellus Shale Drilling? http://stateimpact.npr.org/pennsylvania/2011/09/12/can-pennsylvanias-state-forests-survive-additional-marcellus-shale-drilling/. Accessed 5 June 2014

StateImpact Pennsylvania (2013) Economists Question Corbett's Marcellus Shale jobs claims. https://stateimpact.npr.org/pennsylvania/2013/11/06/economists-question-corbetts-marcellus-shale-jobs-claims/. Accessed 5 June 2014

StateImpact Pennsylvania (2014) Berks County gas-to-liquids plant gets preliminary approval. http://stateimpact.npr.org/pennsylvania/2014/03/14/gas-to-liquids-plant-gets-preliminary-approval-in-berks-county/. Accessed 6 June 2014

U.S. Census Bureau (2010) County business patterns: 2008. U.S. Department of Commerce, Washington, DC

Weinstein, A.L., and M.D. Partridge. 2011. The Economic Value of Shale Natural Gas in Ohio. Columbus, OH: College of Food, Agricultural, and Environmental Sciences, The Ohio State University.

Ward M, Kelsey TW (2011) Local business impacts of Marcellus Shale development: the experience in Bradford and in Washington Counties, 2010. The Pennsylvania State University, Penn State Extension, University Park

Eagle Ford and the State of Texas

Thomas Tunstall

Abstract The Eagle Ford Shale formation has had wide ranging effects in South Texas. For the larger 20-county area, Eagle Ford Shale activity generated over $61 billion in economic impact and supported 116,000 jobs in 2012. These impacts are being felt across some of the traditionally poorest counties in the state of Texas. The build-out of supporting infrastructure such as rail and pipelines has been substantial. There have also been strains on other types of infrastructure such as roads, water and wastewater treatment, and housing. The implications of unconventional shale oil and gas development also have global implications, such as the prospect for exporting natural gas from the United States. Yet, community sustainability remains a key concern for local leaders. In order to assuage concerns about the prospects for an eventual slowdown, community leaders are looking for ways to build high-quality infrastructure and diversify their economies.

Introduction

The Eagle Ford Shale now ranks as the largest single oil and gas development in the world based on overall capital expenditures to date (Dittrick 2012). Wood Mackenzie Ltd. recently calculated that oil and gas companies would spend $28 billion in the South Texas Eagle Ford play during 2013 and estimates for capital spending in 2014 are similarly high. In 2012, many infrastructure projects had commenced or completed construction, including multimillion dollar oil and gas operation centers, pipelines, terminals, and processing plants. Research at the University of Texas at San Antonio's Institute for Economic Development estimated that close to $19 billion was spent on capital expenditures in 2012 (Tunstall et al. 2013).

T. Tunstall (✉)
University of Texas at San Antonio, 501 W. Cesar Chavez Blvd, San Antonio, TX 78207, USA
e-mail: thomas.tunstall@utsa.edu

W.E. Hefley, Y. Wang (eds.), *Economics of Unconventional Shale Gas Development*, Natural Resource Management and Policy 45,
DOI 10.1007/978-3-319-11499-6_6

In May 2012, UTSA released *Economic Impact of the Eagle Ford Shale*, which focused on production, drilling, and related activities (Tunstall et al. 2012). The 2013 study (Tunstall et al. 2013) was adjusted to focus specifically on the impacts of 14 producing counties that are the most active in the Eagle Ford Shale development area: Atascosa, Bee, DeWitt, Dimmit, Frio, Gonzales, Karnes, La Salle, Live Oak, Maverick, McMullen, Webb, Wilson, and Zavala. In addition, significant activity beyond exploration and drilling is occurring in six adjacent counties that are included in the analysis: Bexar, Jim Wells, Nueces, San Patricio, Uvalde, and Victoria. The counties are highlighted in a map on next page. Other counties typically associated with the Eagle Ford include Brazos, Burleson, Edwards, Fayette, Houston, Lavaca, Lee, Leon, Milam, and Wood were not included in the scope of the research due to relatively low levels of drilling activity (Fig. 1).

For the 14 producing counties, the 2012 economic impact was estimated to be over \$46 billion, supporting 86,000 jobs. For the larger 20-county area, Eagle Ford Shale activity generated over \$61 billion in economic impact and supported 116,000 jobs in 2012. Looking ahead to 2022, the 14-county area is expected to generate approximately \$62 billion in economic impact and support over 89,000 jobs. In the 20-county area, the economic impact in 2022 is projected to be over \$89 billion, supporting 127,000 jobs.

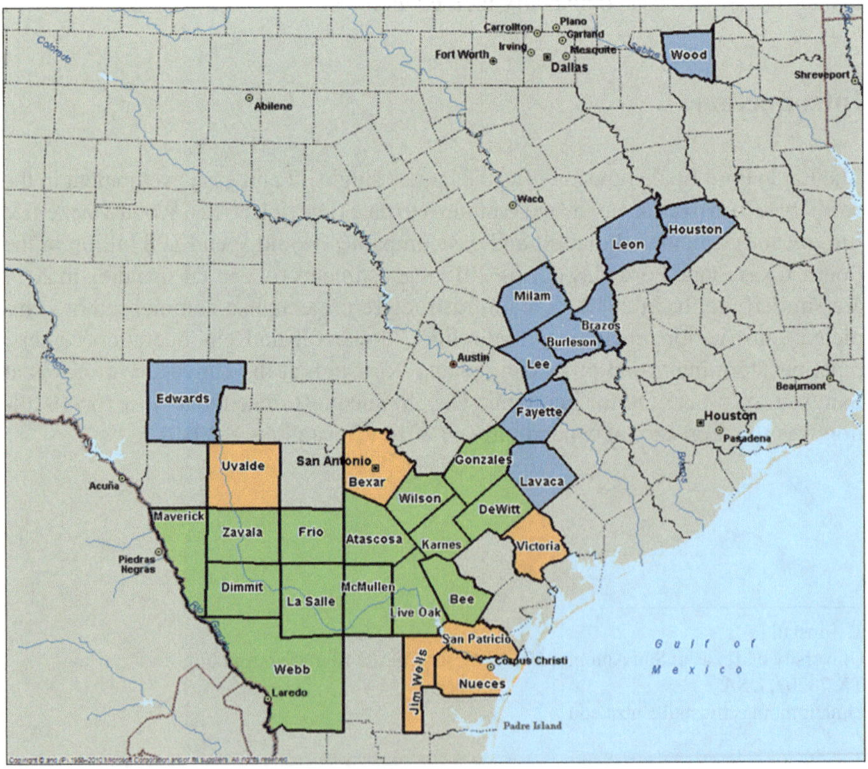

Fig. 1 Eagle Ford Shale Study Area. *Green*: Producing Counties; *Orange*: Adjacent Counties; *Blue*: Counties typically associated with Eagle Ford Shale, but not included in the scope of study

Background

Almost immediately, it is clear that the Eagle Ford Shale presents researchers with unique characteristics worthy of examination. While several peer-reviewed economic development studies have been conducted on hydraulic fracturing for natural gas to date (Blumsack 2011; Christopherson and Rightor 2011; Considine et al. 2011; Kinnaman 2011; Weber 2012), far fewer (if any) have been conducted on unconventional oil plays. Usually, shale fields contain predominantly one or the other resource – either natural gas or oil. The Eagle Ford Shale, however, contains large quantities of crude oil, natural gas liquids, and natural gas. Daily oil production is over 650,000 barrels per day, and natural gas production runs at over 2.6 billion cubic feet per day. This has buffered South Texas from the significant price difference that exists between natural gas and oil. For example, when natural gas prices hit lows of $2 per thousand cubic feet (mcf) in 2012 (down from $8 to $12 mcf in previous years), the slowdown in activity that occurred in predominantly natural gas shale fields such as the Barnett and Haynesville did not occur in the Eagle Ford, because oil prices remained relatively high (between $77 and $109 per barrel). Energy producers in the Eagle Ford simply shifted their drilling activity from natural gas to oil, and production activity continued unabated.

The Eagle Ford (as well as other shale fields in Texas) also benefits from the fact that job growth from oil and natural gas extraction tends to be concentrated in states like Texas where energy companies are headquartered. Such jobs include engineers, corporate managers, and consultants (Rumbach 2011). Texas has a long history of oil and gas exploration, including regulation by the Railroad Commission and a system of established, well-defined mineral rights. These factors are more likely to offset resource curse effects, which postulate that countries or other jurisdictions rich in natural resources often end up worse off economically than those without natural resource abundance (Sachs and Warner 1995). Further, other states or countries that have experienced the resource curse have typically simply exported natural resources with minimal downstream processing (Ross 1999). Texas, by contrast, has more refineries than any other US state and, as a result, generates many additional jobs beyond primary oil and gas extraction.

Nonetheless, community leaders in the Eagle Ford region face significant challenges, as some externalities have not been adequately addressed. These include road infrastructure, police and fire responders, and healthcare capacity. Despite the fact that city and county sales taxes and property tax collections are on the rise, they fall far short of the budget required to replace, repair, or upgrade all of these needs.

Sustainable Community Development

Economic development can be an uneven process, and there is no guarantee that any given community will prosper over the long term. As an example, if we look back at the past 150 years or so, it might surprise us to learn that over 200 ghost towns have evolved in Texas. These towns had growing populations typically from the 1850s until the early 1900s and then saw a significant decline. Some have become completely abandoned. In several cases, a fall in demand for natural resources from a given geography was the proximate cause for the town population to dwindle, though other reasons include highway or railroad bypass, drought, relocation of county seats, creation of man-made lakes, or the widespread consolidation of agriculture that occurred from the 1930s to the 1970s due to mechanization (Baker 2003).

Not so long ago, many communities in South Texas probably had concerns about becoming the next ghost town as well. But then without warning, unconventional oil and gas exploration techniques changed the landscape entirely. Now, local residents are faced with issues regarding sudden resource wealth. Yet, as Texans know probably better than anyone else, booms will sooner or later lead to slowdowns, if not outright busts. The Permian Basin area in West Texas, for example, which is heavily dependent on oil and gas production, has seen ups and downs in its economy related to the price of crude oil for decades.

With a wealth of such cautionary tales, many Eagle Ford Shale communities are working to ensure sustainability based on job creation in a variety of diversified industries, good quality of life, and stewardship of the environment, which are the defining components of current economic development theory (Portney 2013). In addition, the literature suggests that third-wave strategies should be employed to address high-quality physical infrastructure and workforce development (Osgood et al. 2012).

The situation in South Texas is rife with challenges as a result of the shale oil and gas boom. For example, there is clearly the potential for crowding out effects to impact other industries. Restaurants and retail stores have reported difficulty hiring and now offer signing bonuses or have resorted to paying workers twice the minimum wage. School districts and city offices are losing employees to the energy industry. Housing is in short supply, rents have doubled or tripled, and most hotels are regularly sold out. Given these circumstances, it would be very unlikely for a company unrelated to the oil and gas industry (or its support) to consider locating in the area.

Some possible approaches to mitigating impacts from the resource curse have been proffered in the literature (Stevens 2003). These include:

Decrease Production Rates – Essentially argues for undertaking slower development of natural resources. This gives the local economy and society more of an opportunity to adjust, which contrasts with the sudden surge that the Eagle Ford counties (and many others) have experienced. Under this scenario, revenue management for communities becomes easier and crowding out effects are likely to be lessened. The problem, of course, is the difficulty involved in persuading exploration and production companies, landowners, and other suppliers to slow development, particularly with the unpredictable nature of future commodity prices.

Diversification – Clearly, communities can ease boom and bust cycles if they are able to diversify. However, in the midst of a natural resource upswing, it is not typically feasible to attract new industry. Existing infrastructure is already under stress, housing is in short supply, and existing workforces may nearly tap out.

The most communities can do in the middle of a natural resource boom is to undertake planning and then initiate those plans when activity slows down.

Revenue Sterilization – Local governments can moderate aggregate demand and inflation by resisting pressure to spend new tax revenues immediately and instead accumulating budget surpluses.

Stabilization Funds – Diverting tax revenues in order to neutralize the impact of large revenue windfall inflow has been used by many states and countries. Severance taxes in Texas generated from oil and gas production are largely diverted into the Economic Stabilization Fund (also known as the Rainy Day Fund) administered by the Comptroller's Office and the State Legislature. Along those same lines, North Dakota has established a Legacy Trust Fund to better manage the production revenues coming from the Bakken Shale. Norway's Sovereign Wealth Fund also serves a similar function.

Investment Policy – Government can encourage economic diversification and infrastructure development. A large portion of the taxes collected from oil and gas activity in Texas are managed by the state, not at the regional or county level. These include the bulk of the sales taxes collected, as well as severance (oil and gas production) taxes. While sales tax collections at the city and county level have increased significantly in the Eagle Ford area since drilling began, the new revenues are dwarfed by the cost of repairing or replacing existing roadways. County roads and farm-to-market road can range in cost from a quarter to half a million dollars per mile. State highway-grade roads can cost a million dollars or more per mile. Texas recently allocated $225 million for critical repairs to damaged roads in booming areas that include Eagle Ford, as well as the West Texas Permian Basin area. In addition, another $1.2 billion was added to the state's overall transportation funding in 2013 from the Economic Stabilization Fund, which has been bolstered significantly by oil and gas revenues in recent years.

In the case of Texas, at least one of the approaches outlined above is unrealistic (decrease production rates) and others have been implemented (stabilization funds, investment policy). But certainly, diversification will be an important strategy for rural communities in order to both avoid overreliance on the oil and gas industry and move away from dependence on government-subsidized agriculture. More generally, rural communities must develop new industries with sustainable competitive advantages (Deller and Chicoine 1989; Atkinson 2004).

Educational Attainment

Many of the counties in South Texas have been among the poorest in the state, if not the country (Fig. 2). With the exceptions of McMullen, Wilson, and DeWitt Counties, all of those in the study area had poverty rates above the average for Texas, as can be seen in the figure and table below. Eagle Ford Shale exploration and production hold the potential to transform the region if local leaders can seize the opportunity (Table 1).

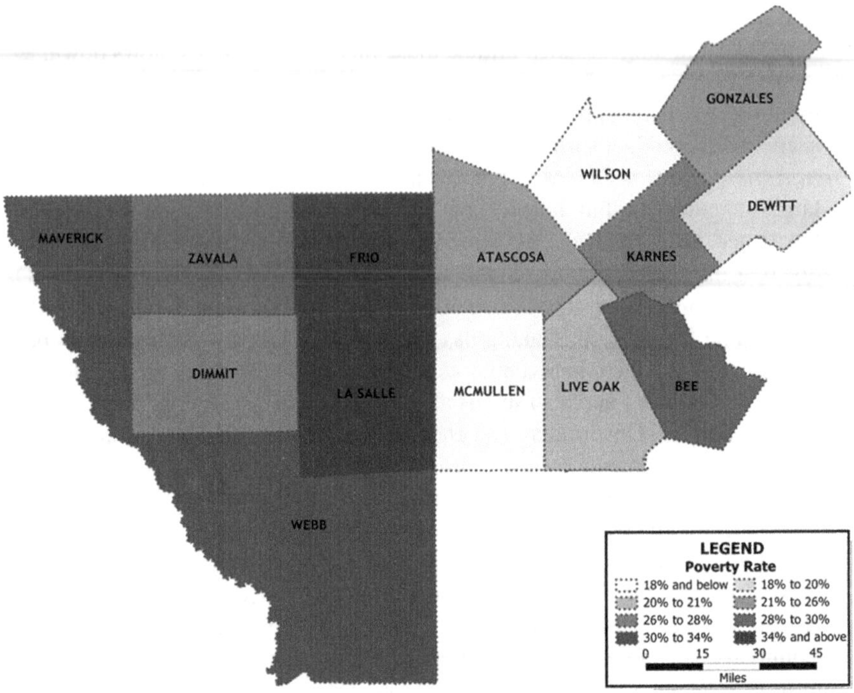

Source: USDA

Fig. 2 Poverty rates in 14 producing counties in the Eagle Ford in South Texas – 2011 (Source: USDA)

Table 1 14-county poverty rates – 2011

County	Poverty rate
La Salle	36.3
Frio	34.6
Webb	32.1
Maverick	31.2
Bee	29.6
Zavala	28.7
Dimmit	26.8
Karnes	26.7
Gonzales	23.4
Atascosa	21.8
Live Oak	20.2
Texas	*18.5*
DeWitt	18.1
Wilson	12.4
McMullen	10.4

Source: USDA

Table 2 Educational attainment 2007–2011

County	Less than high school	High school degree only	Some college	College degree
La Salle	47.2	26.6	20.2	6.0
Maverick	43.9	23.5	20.1	12.5
Dimmit	42.1	30.6	16.1	11.2
Zavala	41.4	21.3	27.6	9.6
Webb	36.4	21.3	25.2	17.1
Frio	36.2	32.3	23.2	8.3
Gonzales	31.0	33.2	22.9	12.8
Karnes	30.6	35.1	24.2	10.1
Bee	28.8	31.3	31.2	8.7
Atascosa	24.8	37.0	26.1	12.1
DeWitt	24.6	36.3	26.9	12.3
Live Oak	22.3	35.3	28.8	13.6
McMullen	21.6	47.0	22.8	8.6
Wilson	15.4	34.5	31.3	18.8
Texas	*19.6*	*25.7*	*28.7*	*26.1*
United States	*14.6*	*28.6*	*28.6*	*28.2*

Source: US Census American Community Survey

The special challenges that rural communities face include lower than average educational attainment. South Texas is no exception. Table 2 highlights the high numbers of residents who have less than a high school education, which translates into lower educational attainment overall. This in turn inhibits economic development in South Texas. Local higher educational institutions have developed programs to address many of the emerging needs of the oil and gas industry, which could also improve educational achievement over time along with prospects for the local workforce.

Opportunities for Economic Diversification

For the rural communities in South Texas, the potential options for employment growth include higher-margin agricultural products such as olives and olive oil processing, spinach and other food processing, geothermal energy, tourism, hunting, outdoor recreation, water recycling/desalination, and wine/beer making. The prospects for some of these industries are outlined below.

Olives and Olive Oil Processing – The United States imports nearly 300,000 tons of olive oil annually and produces only about 12,000 tons. Production of olive oil in Texas has risen from nothing in 2002 to approximately 54 tons in 2012. The number of olive trees in Central and South Texas is rising rapidly, from approximately 250,000 in 2012 to an anticipated 1,500,000 in 2013. There are four olive oil pressing plants in Texas, with others planned in the future. Olives and olive oil are a higher-margin agricultural growth industry, and olive oil consumption in the United States has been increasing because of research that consistently demonstrates the health benefits of a Mediterranean diet.

Geothermal – Alternative energy sources such as geothermal, which have much smaller carbon footprints than fossil fuel, will become increasingly attractive. Geothermal is more reliable than wind or solar, as plants operate 24/7. There are several viable sites for geothermal in South Texas, which presents a growth opportunity for a green energy source. The geothermal industry employs several types of high-skill positions, many of which are very similar to job categories employed by the oil and gas industry. As such, this industry could provide workers with a transition industry to migrate toward in the event of a slowdown in oil and gas production.

Water Recycling and Reclamation – Given the impacts of the current drought combined with projected substantial population increases for Texas, opportunities to provide water from nontraditional sources, such as recycling and desalination, are likely to increase. Such water projects are applicable to both potable and non-potable uses. Water recycling and desalination can decrease the diversion of freshwater from sensitive ecosystems, as well as lakes and aquifers in Texas. Here again, many job openings will require high-skilled technical experts who can often work remotely. Water is a particularly critical issue for growth, as evidenced by the fact that the State Legislature of Texas recently proposed and the citizens approved a constitutional amendment to authorize $2 billion for reservoir, wells, and conservation projects.

Tourism – Texas has been a strong draw for tourists and other types of visitors. Estimates are that in 2012, the 14 counties generated over $1 billion in visitor spending (Klein 2013). Many historic sites in South Texas relating to Texas Independence, Spanish settlements, and the early days of cowboys are a few examples that local communities could capitalize upon.

In addition to the above industry diversification examples, there are additional opportunities based on emerging trends. If a robust broadband infrastructure can be put in place, there are prospects for telemedicine, distance learning, re-shoring of jobs previously outsourced overseas, and attracting knowledge workers who prefer the lifestyle associated with smaller communities. Rural communities have traditionally lagged metropolitan area in terms of income, which has been an impediment to job growth. Improvements in information and communication technologies (ICT) in rural areas would be expected to improve the prospects for new residents to earn an economic livelihood there (Albrecht 2012).

Educational opportunities in rural areas tend to be limited. As a result, emerging distance learning opportunities present real prospects for sustainability. Research has demonstrated that rural areas without access to institutions of higher education have a much harder time attracting educated workers and building their human capital stock (Winters 2011). In addition, re-shoring of many previously outsourced job functions or expanding the US trade surplus in services could become more feasible in rural areas with improved ICT. And finally, with increased cost pressures likely as a result of healthcare reform, telemedicine offers significant opportunities to expand delivery networks and increase efficiency to nonmetro areas.

Sales Tax Revenue

Property tax collections provide the largest source of revenue that local governments have available for providing education, transportation, and law enforcement. When local governments plan budgets, property tax revenues are considered to be a stable monetary source. However, increases in these revenues can take up to a year following development to take place.

One current issue faced by communities in the Eagle Ford Shale region regarding property tax revenue is that property taxes are not generally collected from recreational vehicle (RV) occupants in the counties. Similarly, RV park owners likely do not pay proportionate property taxes commensurate with the strain that the residents place on local infrastructure and municipal services.

While housing developments are currently under construction in various locations throughout Eagle Ford, there remains a substantial lag time for collecting property tax revenue on these homes. Despite these challenges, property taxes are expected to increase significantly as residents continue relocating to the 14-county area causing property values to rise.

Sales taxes collected in the Eagle Ford counties have increased dramatically between 2010 and 2012. This tax, imposed on all retail sales, leases, and taxable services, has created a new source of revenue for local communities (Fig. 3).

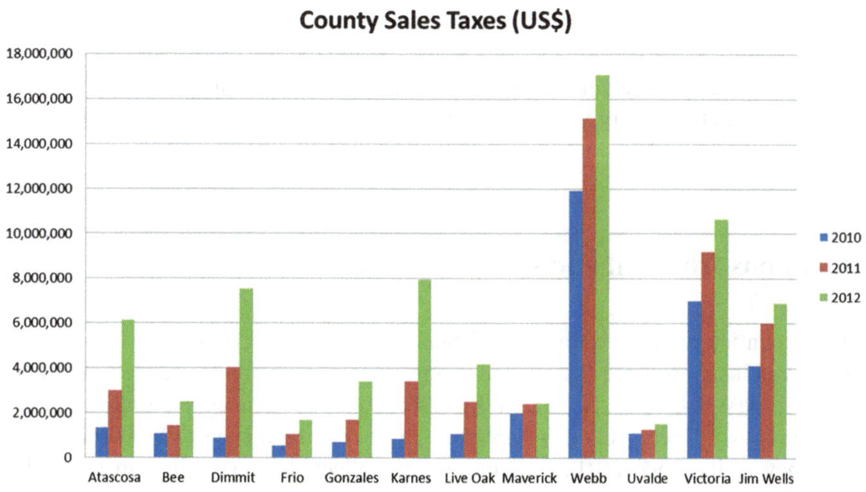

Fig. 3 Tax Collections for Counties in the Eagle Ford

County Transportation and Infrastructure

Road Damage

Poor road conditions and increased traffic have become a cause of concern for many residents in the Eagle Ford Shale. Hundreds of 18-wheel vehicles regularly run up and down many county and farm-to-market roads causing a tremendous amount of deterioration on South Texas highways. Everyday, trucks hauling drilling rigs, oil field equipment, chemicals, or wastewater travel on state highways and smaller county and ranch roads built for light vehicle or tractor traffic.

For both the state and the counties, there is no ideal source of funding in place to mitigate road repair costs. State law does not give counties the authority to mandate additional road repair fees on companies that already have drilling permits from the state. While the Texas Rainy Day Fund (formally known as the Economic Stabilization Fund) is expected to reach as much as $12–14 billion by the start of the 2015 legislative session – funded largely with oil and gas severance taxes from areas such as the Eagle Ford – there is no formal mechanism to ensure that the costs associated with roads are addressed by the severance tax collections.

Estimated Costs for Repair

The Texas Department of Transportation has not fully calculated the potential long-term road maintenance costs associated with drilling in Eagle Ford, but early estimates indicate that roadways presently require roughly $2 billion total in maintenance – $1 billion for damage done to state highways and $1 billion for damage done to municipal and county roads (Hiller 2012).

In DeWitt County alone, Naismith Engineering Inc. of Corpus Christi estimated that the county's nearly 400 miles of roadway would require more than $400 million in construction and maintenance over the next 20 years. This amounts to $350 million more than the county would have previously allocated for roads and represents millions more than the county received from the state in 2012.

Donations from Oil and Gas Companies

Many counties within the Eagle Ford Shale play receive "donations" from the oil and gas industry or have "gentleman's agreements" to provide materials for road repair. In DeWitt County, two major drilling companies – Pioneer Natural Resources and Petrohawk Energy Corp. – agreed to pay a road repair fee of $8,000 for each well they drill. Since 2010, the county has collected $1.6 million in fees, plus another $2 million through voluntary agreements.

State transportation crews and county road departments are working to improve the most dangerous sections – roads narrowed to less than 22 feet. Counties have tried leveraging a combination of property tax revenue and voluntary fees paid by oil and gas companies for these maintenance projects, but state and county officials across the region are looking for a more systematic solution to address road issues.

Implications for the Political Landscape

Not surprisingly, infrastructure needs in South Texas have become a significant topic of discussion in the past couple of years. Roads in the Eagle Ford Shale, for example, are under intense pressure from the voluminous truck traffic that now runs up and down South Texas highways – literally hundreds of trips per day on many of them. What is becoming apparent is that there is a disconnect in the Texas political economy between how tax revenues are generated and how roads are then funded. Given TxDOT's recent announcement that approximately 83 miles of FM roads have been slated to be returned to gravel (66 miles of them in the Eagle Ford area), it's worthwhile to examine road funding mechanisms in Texas, as well as larger issues affecting the political landscape.

Starting with the state gas tax at the pump, which is a total of 38.4 cents, immediately, 18.4 cents goes directly to the federal government, which leaves 20 cents for state use. However, 5 cents of that goes to public education. Only the remaining 15 cents is used to fund TxDOT projects directly. Texas motor vehicle fuel sales taxes are flat taxes that have not been raised since 1991 and are not adjusted for inflation.

The unprecedented activity on the roads in the Eagle Ford Shale area is having an impact that is overwhelming traditional highway funding sources. As an example, it takes nearly 1,200 truck trips (equivalent to 8 million cars) to complete a single oil or gas well. Another 350 or so are estimated to be required for annual production.

Potential Funding Sources for Roads

Sales taxes in Texas have a statutory maximum rate of 8.25 %. Of that total, 6.25 % goes to the state. Cities, counties, transportation authorities, and economic development corporations can add up to an additional 2 % to their sales tax rates. Some counties charge no sales tax, such as McMullen County, so the maximum rate there is 6.25 %. Since city and county sales taxes in the Eagle Ford Shale area have increased significantly starting around 2010, it might seem to make sense for these entities to pick up the tab for increased road wear. In some cases, for example, county tax increases jumped between 300 % and 500 % in a single year. Unfortunately, the local sales tax collections pale in comparison to the cost of building roads.

County roads, for example, typically cost around $250,000 per mile to build. Farm-to-market and farm-to-ranch (FM) roads cost twice that – about $500,000 per mile. State highway-grade roads cost in excess of $1 million per mile. When county and FM roads are repaired to their current standard, the cost can be less – about $120,000 per mile – but heavy volumes of truck traffic can tear them back up in less than a year.

Taking the case of a specific county can be instructive: One of the most active counties in terms of Eagle Ford production is Karnes County. In 2010, county sales tax receipts were $837,038. By 2012, that number had risen to $7,961,495 – a huge increase by any measure. And yet, if every dollar of increased county sales tax revenue were applied to roads in the area, Karnes County would be able to build about 28 miles of county roads, 14 miles of FM-grade road, or only 7 miles of state highway-grade road. Clearly, the orders of magnitude for the road impact as a result of oil and gas exploration and production activity are beyond the scope of county budgets.

In fact, some of these severance taxes are being channeled to road projects. During the most recent legislative session, $1.2 billion per year was allocated from the Rainy Day Fund for roads across the state (pending approval by voters in November 2014). In addition, a one-time infusion of $225 million was allocated for road systems in South and West Texas areas affected by oil and gas production. And just this month, TxDOT announced that it had identified another $250 million from vehicle registration fees.

It is becoming clear that several aspects related to the costs of shale oil and gas production (roads in particular) will not necessarily be remedied by current tax revenue mechanisms. As such, any chance for a more permanent solution will be up to the Texas Legislature. And yet, the prospects for legislative remedies will be more difficult than in the past because of the way the population in Texas has shifted over the past century.

Years ago, Texas was a predominantly rural state. Populations of cities and counties in the late 1800s and early 1900s were much more evenly distributed back then. If we look back to the 1860s, we would note that nearly 60 % of the US workforce consisted of farmers. In 1900, it was still about 40 % of all workers. Now of course, only 2 % of the US workforce is employed in agriculture. As a result, fewer people live in rural areas, and the fastest growing geographies in Texas are now the larger cities. This shift in the distribution of the state's population has implications important to the Eagle Ford Shale area (and West Texas as well) in terms of legislative representation.

In 1890, for example, Gonzales County had around 18,000 people living there. San Antonio had a little over 37,000 and Bexar County had just fewer than 50,000 people. By 2000, San Antonio had over one million residents and Bexar County boasted over 1.3 million – increases of 2,500 % or more. Yet, in 2000, Gonzales County still had about the same number of people as in 1890.

This is indicative of the growth occurring in the larger cities like San Antonio, Houston, Dallas-Fort Worth, and Austin. And yet what often goes unnoticed is that both Texas Senate and House seats are apportioned by population. Unlike the US

Senate, where every geography (state) has retained two votes since statehood, the Texas Senate is population proportional. So as communities in South and West Texas lose ground to the larger cities in terms of population growth, they lose not only House but also Senate seats as well.

In the Eagle Ford Shale geography of the 20 counties that UTSA's Institute for Economic Development has been studying over the past few years, another example is instructive. In 1900, Bexar County, for example, only contained 31 % of the population, which meant that almost 70 % of people lived in the other parts of the Eagle Ford area. By 2010, however, Bexar County's share of the 20-county Eagle Ford Shale population had doubled to 61 %. With that growth, comes a greater political voice in terms of more State Representatives and Senators for cities like San Antonio and less for rural counties in the Eagle Ford.

Some of the most dramatic population shifts have occurred since the end of World War II, when agricultural mechanization began to systematically decrease the number of people employed on farms. From 1950 to 2010, DeWitt, Dimmit, Gonzales, Karnes, La Salle, and McMullen Counties all lost between 6 % and 40 % of their population. In that same period, San Antonio and Bexar County increased over 200 %. Many counties in West Texas now being impacted by the Cline and other shale discoveries have seen similar population decreases since the 1950s.

The reality of Texas politics is that all parts of the state are in constant competition for the limited highway funding available. Dallas-Fort Worth, Austin, and Houston, for example, have their own issues with regard to roads. While South and West Texas are seeing the impacts in the form of road deterioration from large numbers of 18-wheelers, the big cities struggle with increasing congestion because of rapidly growing populations. Both groups make a good case for increased highway funding, but the more populated cities and counties have a much greater political voice than in the past simply because they have more State Senators and Representatives. Given the shift in political clout to the larger cities in Texas, it will be important for the communities in South and West Texas to work together to make their case to the Texas Legislature.

Billions of dollars in severance taxes are being generated from exploration and production activity in South and West Texas. Rural Texas also provides important agricultural products, wind energy, hunting, recreation, and tourism, among others. So in addition to serving the needs of the large urban areas, Texas legislators should take care to make sure rural Texas is served also.

Of course, beyond legislative remedies, rural communities across Texas must seize the opportunity to reinvent themselves. The predominant family farm system that was characteristic of rural Texas in the late nineteenth and early twentieth century has changed because of technological progress that requires fewer people in traditional agriculture. But the population of Texas is growing (nearly 47 million people estimated by 2060 – up from 26 million currently), and this trend presents opportunities for rural areas to grow also if they can establish an infrastructure that attracts new residents, visitors, and businesses. The chance to do just that is now possible due to recent shale oil and gas wealth.

Railroads and Eagle Ford

The success of railway in Eagle Ford is strongly linked with the lack of pipeline infrastructure in place when the play was initially developed. Railroad has provided producers and suppliers in Eagle Ford with an efficient and timely means for transporting their product to refiners.

Likewise, railroad has been integral in supplying field service operators with raw materials such as frac sand, gravel for well pads, and lumber, all essential to the hydrocarbon extraction process. Railroads have also been instrumental in transporting countless miles of pipeline that will be buried throughout the shale play as midstream operators seek to find long-term solutions to the logistical problems presented by increased Eagle Ford production.

Union Pacific (Fig. 4) operates 6,319 miles of track in Texas and an overwhelming majority of the rail lines servicing the Eagle Ford play. Due to the development of Eagle Ford and other fields, carloads of crushed stone, gravel, and sand moved annually increased roughly 37 % across the system between 2009 and 2012. Lumber and wood carloads per year increased 20 % during the same time period. The number of railcars terminated in Texas also increased by 21 % since 2009, surpassing volumes prior to the Great Recession. The total number of railcars originated in Texas increased by 11.5 %, also exceeding prerecession volumes.

Evidence of the railroad logistics boom can also be seen in the current demand for tanker railcars used to move crude oil to refineries and pipeline terminals. Three of the largest railcar manufacturers – Union Tank Car, American Railcar Industries, and Trinity Industries – are struggling to keep up with demand, running their fabrication

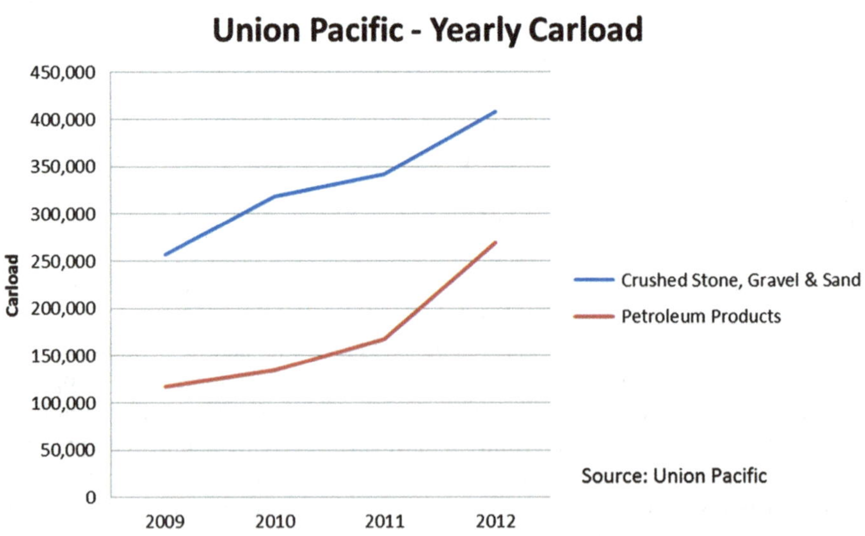

Fig. 4 Changes in Rail Traffic in the Eagle Ford

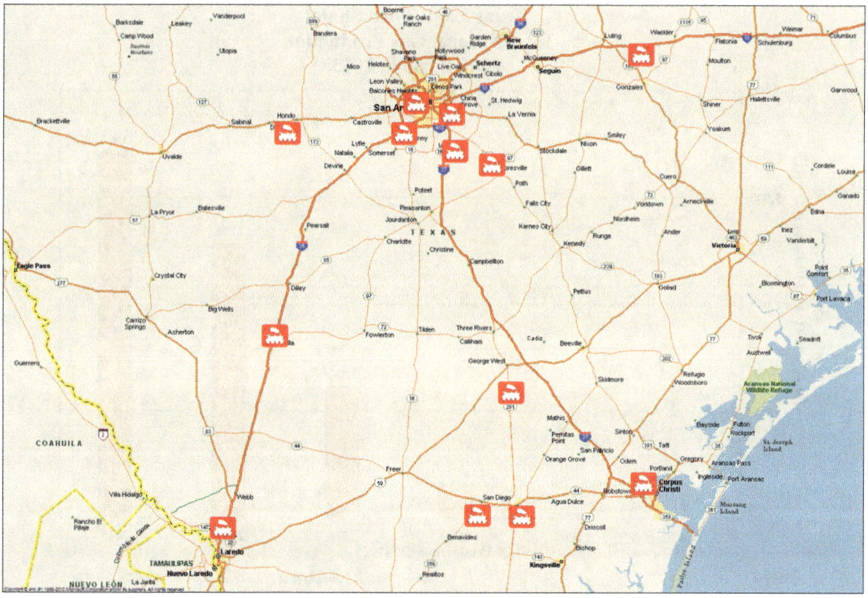

Fig. 5 New or Upgrade Rail Facilities in the Eagle Ford

facilities at full capacity. In some cases, these companies are performing conversions of wind farm tower factories to meet needs. Crude oil can readily be transported via rail on cars that can hold up to 725 barrels each, strung together up to 100 at a time. By the end of the 3rd quarter of 2012, the railcar manufacturing industry's backlog was roughly 46,700 (Black 2013).

New railroad projects and expansions have been undertaken in or near major cities associated with Eagle Ford activity such as San Antonio, Corpus Christi, Houston, and others (Fig. 5). But essential to the logistical success of these expansions have been auxiliary railroad interchanges and yards operated by short-line carriers. These specialty carriers support the logistical, terminal, and storage solutions to pipeline, oil field, and other Eagle Ford operations. Many of the interchange yard facilities are located midway between the production sites and major marketplaces, such as port and refining facilities.

Eagle Ford and the Texas Gulf Coast

Eagle Ford's impact on the Gulf Coast of Texas has yet to be fully calculated and understood. Nonetheless, it is clear from the build-out of pipeline and midstream infrastructure that the energy industry is making industrial use of Eagle Ford natural gas and its by-products.

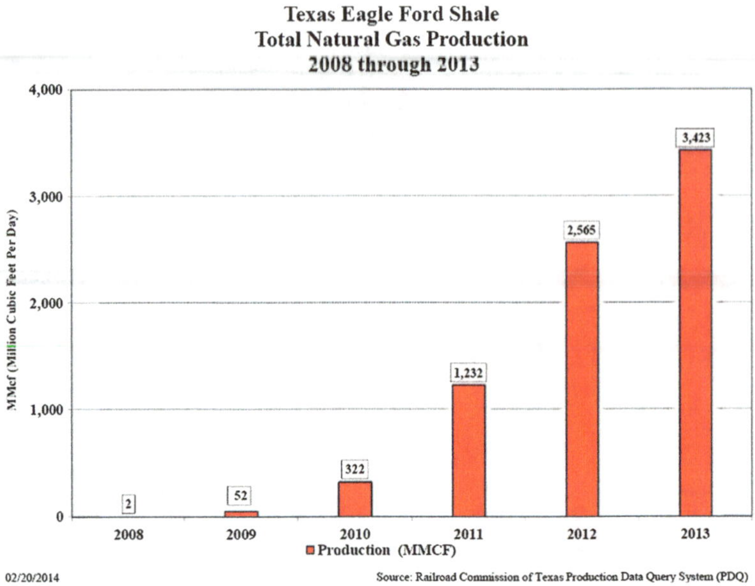

Fig. 6

Early development of the Eagle Ford play centered largely on the production of natural gas. However, wellhead prices for natural gas dropped steadily beginning in July of 2011 reaching record lows in April of 2012 and finally recovering somewhat in 2013 (Fig. 6). Natural gas production in 2013 reached 3.4 billion cubic feet per day or about 1.2 trillion cubic feet annually.

Lower costs for natural gas as a result of abundant supplies made possible by horizontal drilling and hydraulic fracturing technology have spurred new plans for investments in energy infrastructure. The driving force behind these new investments has largely been the availability of low cost of natural gas liquids, which are removed from natural gas during cryogenic processing. Ethane and propane are key NGLs used in industrial application and are essential to the production of numerous products including plastics, rubber, fertilizer, adhesives, and specialty additives (Fig. 7).

Natural gas liquids are extracted from gas wells in much the same way that a dry well produces methane gas. Eagle Ford is considered to be significantly "wetter" compared to other gas shale plays, generating roughly 4.0–9.0 gallons of NGLs per thousand cubic feet.

According to Economist Jesse Thompson of the Federal Reserve Bank of Dallas, the US petrochemical industry relies heavily on NGLs to produce ethylene. Other areas of the world (except the Middle East) rely heavily on Naptha, the price of which is tied directly to oil. Through the divergence of oil and natural gas prices, petrochemical producers in the United States, and especially along the Gulf Coast

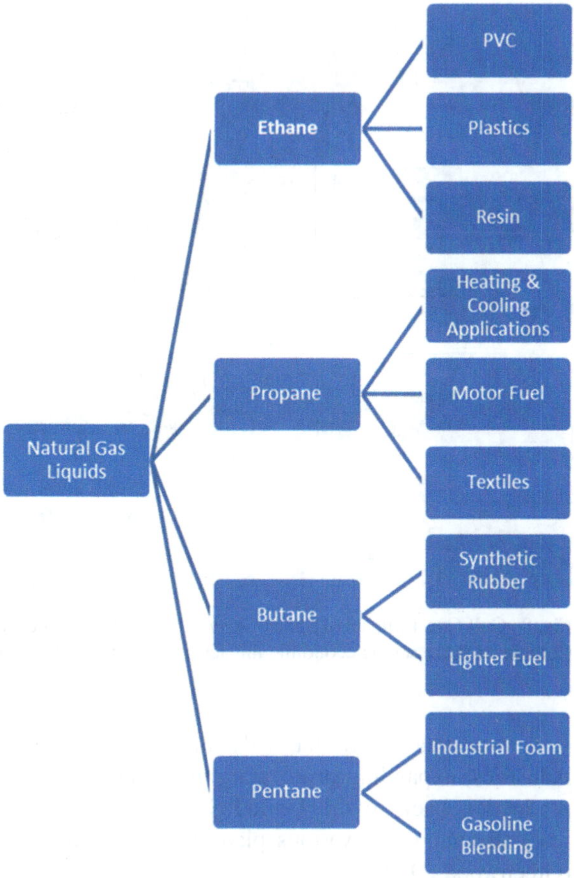

Fig. 7 Breakdown of Natural Gas Liquids. Source: Platts 2012

(Fig. 8), receive a significant cost advantage directly related to the increased supply of natural gas liquids coupled with the commodity's low cost (Thompson 2012).

Low-cost natural gas has also spurred reinvestment in fertilizer production and related facilities. According to The Fertilizer Institute, natural gas represents 70–90 % of the cost for producing nitrogen, which is the heat source of the chemical process used in production. Prior to the shale gas revolution, production of fertilizer in the United States was a stable industry but never able to achieve profit margins that would incentivize new construction of facilities. Because of significant price volatility between 2000 and 2010, operations at domestic fertilizer plants were shuttered and some were moved overseas to the Middle East. However, the profit margins have seen marked improvements due to the recent availability of inexpensive natural gas, which has increased investment related to the fertilizer production (DeWitt 2012).

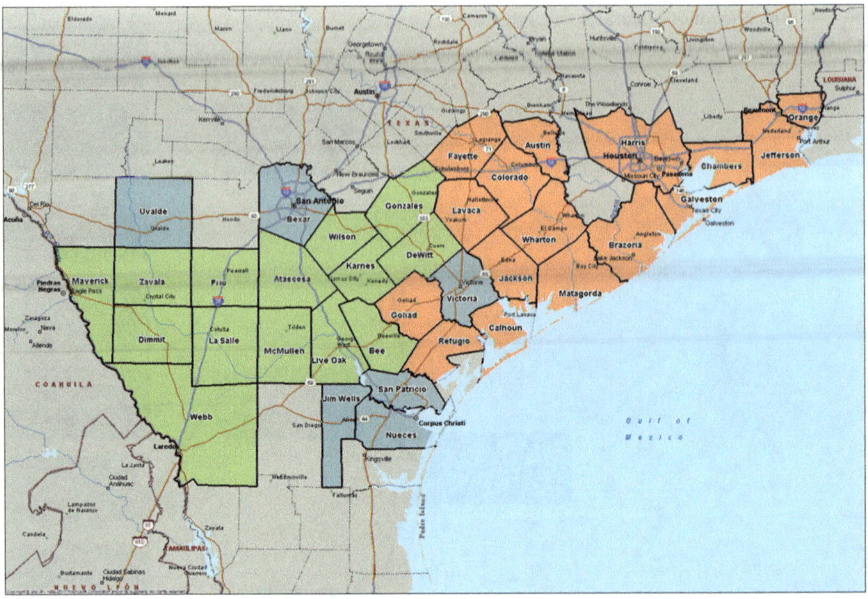

Fig. 8 Gulf Coast Impact of Shale. *Green*: Producing Counties; *Blue*: Adjacent Counties; *Orange*: Counties that are seeing economic impact from the Eagle Ford and other shale fields

Low-cost natural gas has had similar effects on the rubber and tire manufacturing industries, expanding profit margins for domestic producers and luring overseas production back to the United States. Continental, Bridgestone, and Michelin have all announced various plans for expansion in the vicinities of shale fields (Greenwood 2012).

Port of Corpus Christi

Along the Port of Corpus Christi's inner harbor, M&G group, an Italian resin manufacturing firm, plans to begin construction of a 1 million ton per year polyethylene terephthalate (PET) plant and a 1.2 million ton per year purified terephthalate acid (PTA) plant. Both chemicals are manufactured from ethane and are used in the production of resin and packaging materials (Savage 2012).

Both facilities represent a $751 million dollar investment during the 30-month construction period and 250 employees upon completion. In January of 2013, M&G group further solidified their commitment to the construction of the petrochemical plants in Corpus Christi when it signed a $1 billion construction agreement with Chemtex Global. Company officials indicated the construction period would conclude in 2016 given that environmental permits are obtained in a timely manner (Collette 2013).

Although steel manufacturing is not as prevalent along the Texas Gulf Coast as it is in other shale fields such as the Marcellus, opportunities for the development of this industry have been increasing due to natural gas availability. Historically, the steel industry has clustered around the northeast and Midwest and recently experienced a significant amount of growth due to shale gas production. The convenience of the Port of Corpus Christi's location relative to Eagle Ford has helped to spur development in this sector along the Gulf Coast (Casselman and Gold 2012). For example, Tianjin-based steel pipe manufacturer, TPCO, selected a 253-acre site for its Texas mill project. The $1 billion dollar plant, currently under construction outside Gregory, San Patricio County, is expected to produce 500,000 metric tons per year of seamless steel pipes used in oil field applications. The construction phase requires an estimated 300 to 400 contractors, while mill operations will require an initial 300–400 employee workforce (Smith 2012a).

In early February of 2013, it was reported that Occidental Chemical Company filed with the US Environmental Protection Agency to build an ethylene plant capable of producing 1.2 billion pounds of ethylene per year near Ingleside, San Patricio County. The plant would collect NGL feedstock derived from a fractionator to be built adjacent to the OxyChem property. Ethylene produced would be transmitted via pipeline to the adjacent vinyl chloride monomer plant owned by OxyChem. If approved, construction was planned to begin in December of 2014 with production expected in February 2017. The plant would employ 123 individuals according to the application.

Energy analysts and industry executives have come to view natural gas as a desirable feedstock that will generate economic growth in the United States. IHS Vice Chairman Daniel Yergin has indicated that growth in shale gas production will save the United States $100 billion it would otherwise have spent on imported liquefied natural gas (Yergin 2012). Royal Dutch Shell Chief Executive Officer Peter Voser claimed that "it (the United States) can bring manufacturing and petrochemical industries back, and that is where the jobs are. I would find it peculiar if the U.S. does not grab this opportunity" (Gold 2013).

Crude Oil and Eagle Ford

Upward trends in oil production in Eagle Ford have made significant impacts on import patterns as well as Gulf Coast refinery operations. Although it is difficult to assess the amount of oil produced in Eagle Ford that is making its way into these refineries, it is evident from pipeline builds as well as modifications and expansions that many along the Gulf Coast expect to continue intake of the product for the foreseeable future.

The Eagle Ford formation has proven to be a robust source of a variety of hydrocarbons, which enables energy companies operating in the area to adapt to changes in the market. Eagle Ford contains dry gas, heavy wet gas, oil, and condensate. Oil rig operations in Eagle Ford have increased significantly since the summer of 2012 (Fig. 9), when natural gas prices at Henry Hub hit record lows of $2 per thousand cubic feet.

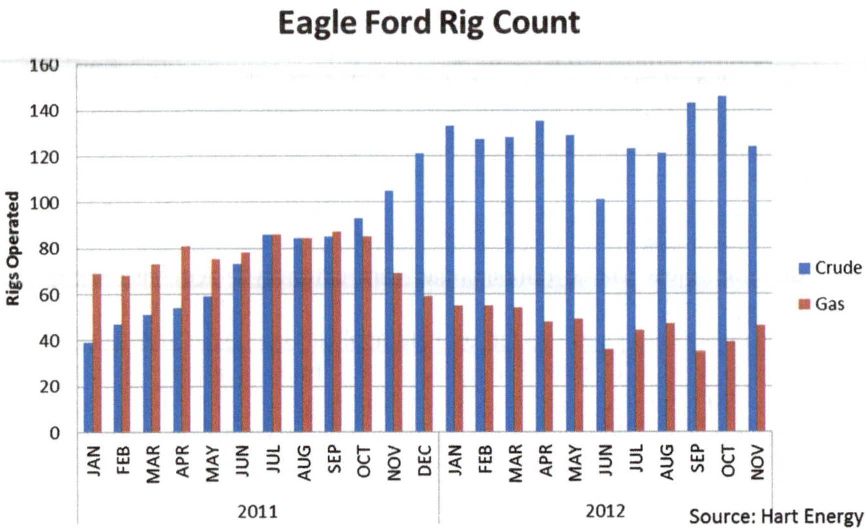

Fig. 9 Oil and Gas Rig Counts in the Eagle Ford

Oil production in Eagle Ford enjoys a distinct advantage over other shale plays because the field is less than 150 miles from refining complexes situated along the Texas Gulf Coast. According to the Energy Information Administration, Gulf Coast crude refinery capacities represent roughly half of the United States' 15.3 million barrel per day total capacity, with Texas alone making up 27 % of the total. In response to recent developments, refiners along the Gulf Coast have begun to reduce the amount of foreign crude oil they import in favor of Eagle Ford and other US-based sources.

According to data collected from the Texas Railroad Commission, oil produced in Eagle Ford has increased from roughly 5.5 million barrels in 2010 (15,149 barrels per day) to more than 250 million barrels in 2013 (688,429 barrels per day; Fig. 10). Likewise, condensate production has increased from approximately 6 million barrels in 2010 (18,784 barrels per day) to over 72 million in 2013 (198,373 barrels per day; Fig. 11).

Refiners in Corpus Christi and Houston stand to benefit greatly from the significant increases in production from Eagle Ford, as well as the low transportation costs. However, until recently, logistical problems related to getting Eagle Ford crude to market have been an impediment. Several pipeline projects associated with Eagle Ford crude transmission were completed in 2012 and 2013. Crude pipeline reversals and extensions have been commonplace in the play as some infrastructure previously existed due to ongoing crude refining operations in Corpus Christi, Three Rivers, and South San Antonio. A number of refineries along the Texas Gulf Coast are using sizeable quantities of Eagle Ford crude in their refineries because overall production of crude in Texas has increased significantly in the past 3 years.

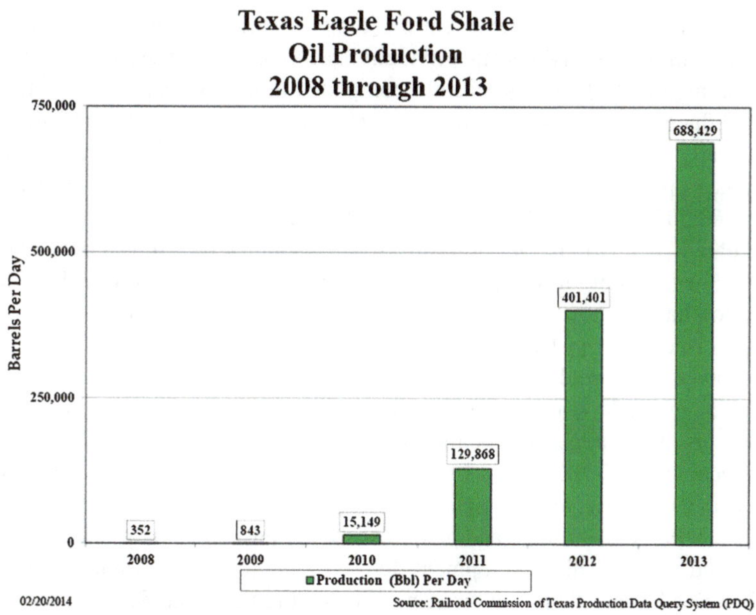

Fig. 10 Oil Production in the Eagle Ford (measured in barrels)

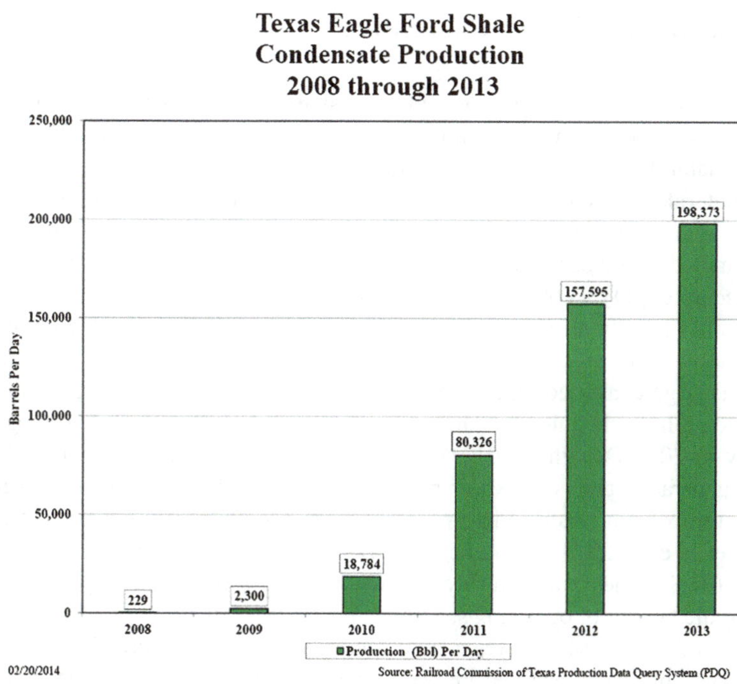

Fig. 11

Other refiners along the Texas Gulf Coast have added capacity and retooled in preparation of increased domestic production. Motiva Enterprises, a joint venture of Saudi Aramco and Shell, has recently completed a massive expansion of the Port Arthur refinery that will allow it to run 600,000 barrels per day of crude, including domestically produced light sweet crude (Dukes 2012). Likewise, Valero continues to enhance operations at the company's nearby Port Arthur refinery where it has added a 60,000 barrel per day hydrocracker (Dukes 2012).

Increased production from domestic shale resources including Eagle Ford has reduced the amount of foreign imports to the Gulf Coast. Eagle Ford crude production has played a significant role in meeting refiner's needs for light sweet crudes. Former Valero CEO Bill Klesse suggested that imports of light sweet crude to the Gulf Coast could cease in 2013. Combined with forecasts regarding crude production in Eagle Ford, as well as other shale plays in West Texas and North Dakota, this in fact appears to be the case.

In December of 2012, Valero Energy secured a permit to ship crude oil from the Texas Gulf Coast to its refinery near Quebec, Canada. A majority of the crude would come from Eagle Ford, displacing what the company previously purchased from Europe and Africa. This development demonstrates the cost advantage associated with increased domestic production and is changing the dynamics of crude import and export in the United States (Vaughan 2012b).

Natural Gas and Power Generation

The abundance of natural gas sourced from shale beds has directly impacted the economics of power generation in the United States. According to the Wall Street Journal, natural gas priced at $3 per thousand cubic feet, down from $8 in 2008, allows natural gas-fired power plant operators in many cases to generate electricity for less than the cost of a coal-fired equivalent (Smith 2012b). This cost differential has prompted a number of closures of coal-fired plants across the United States, estimated at roughly 9,000 megawatts in 2012 alone (Reuters 2013).

Texas has largely mimicked this national trend as evidenced by the number of coal-fired plant closures in recent years. Luminant recently announced plans to idle two of its large coal-fired units in Northeast Texas. City Public Service Energy of San Antonio has scheduled the shutdown of two coal units with an 870-megawatt capacity in 2018 (Vaughan 2012a). Likewise, plans to construct other coal-fired electric generation plants have been put on hold or rescinded. The Las Brisas Energy Center, a petroleum coke-fired power plant project to be located in Corpus Christi, was halted in early 2013. NRG Energy, in December of 2012, chose to drop plans to construct an 800-megawatt coal-fired plant northwest of Houston citing the decrease in economic benefit of operating such a plant (Reuters 2012).

In the wake of these closures, new power generation plants fired by natural gas have opened in recent years or are beginning the permitting and construction process. The Lower Colorado River Authority contracted with Fluor Corporation to construct a 540-megawatt natural gas-fired plant at Horseshoe Bay on Lake

LBJ. Similarly, the South Texas Electric Cooperative, which completed work on a 200-megawatt gas-fired facility near Pearsall in 2010, has entered into a contract with Wärtsilä to construct a 225-megawatt gas-fired plant in Hidalgo County. In neighboring Cameron County, Tenaska Energy has entered into a development and purchase agreement with the Brownsville Public Utilities Board to build an 800-megawatt gas-fired power plant. Projects such as these across the state are significantly reducing emissions and improving air quality in an unexpected fashion. Conventional wisdom assumed that improvement in air quality would come largely from renewable electricity production, not because of increased use of natural gas.

Despite the decrease of coal usage for electrical generation in Texas, the state remains a significant producer of coal as evidenced by the approval of a coal mining operation in Maverick County, near Eagle Pass (Barer 2013). This development demonstrates the growing export economy for coal in the United States because the commodity has become more valuable in the Asia and Europe. According to Energy Information Administration data, 75 % of coal exports were shipped to Asian and European markets in 2012 (EIA 2012b). In the case of Maverick County coal, export was planned for Mexico.

The economics and environmental benefits of natural gas-fired power plants have become so appealing that several electricity providers have shuttered smaller nuclear plant operations due to high overhead costs and the ability to purchase cheap electricity on the open market (Smith 2013).

Increased Use of Natural Gas Vehicles

CNG (compressed natural gas) is composed primarily of methane and is made by compressing natural gas to less than 1 % of its volume. Many companies with fleets now use CNG as a fuel because of its lower cost and reduced emissions. The International Association of Natural Gas Vehicles estimates that there will be more than 65 million natural gas vehicles worldwide within the next 10 years or about 9 % of the world transportation fleets.

The fuel storage tanks on an NGV are thicker and stronger than gasoline or diesel tanks, and there has not been an NGV fuel-tank rupture in at least 2 years in the United States. Natural gas burns cleanly and results in less wear and tear on the engine, thus extending the time between tune-ups and oil changes. According to the American Public Transportation Association, nearly one-fifth of all transit buses were run by compressed natural gas (CNG) or liquid natural gas (LNG) in 2011. Currently, transit buses are the largest users of natural gas for vehicles.

Mass Transportation with CNG Vehicles in Texas

The state of Texas has only recently begun to encourage the development of natural gas fueling stations, passing legislation in 2011 intended to build a market for natural gas vehicle usage. Known as the Clean Transportation Triangle, the program has sought to encourage the development of CNG and LNG fueling stations along the interstate highway corridors between San Antonio, Houston, and Dallas-Fort Worth. The CNG/LNG refueling infrastructure that the state hopes to generate along these major corridors will be publicly accessible – incentivized through a series of grants administered by the Texas Commission on Environmental Quality (TCEQ). Additionally, the TCEQ offers grants to individuals and businesses to convert or replace gasoline- or diesel-fueled vehicles (Garza 2012).

Corpus Christi has an extended history with the use of CNG-powered vehicles that began in the early 1980s with a 30-police vehicle pilot program. In 2010, the city received two grants from the State Energy Conservation Office and the US Department of Energy that allowed it to both construct a new CNG refueling station on city property and convert a total of 26 fleet vehicles (Basich 2011).

The Corpus Christi Regional Transportation Authority (CCRTA) has likewise begun to invest heavily in the use of compressed natural vehicles for their fleet. In March of 2012, the transit authority approved the construction of a $2.1 million CNG fueling station. Likewise, the CCRTA is planning on converting its 81-vehicle fleet of buses and support vehicles to compressed natural gas by 2017. The transit authority estimates it will significantly reduce fuel costs and generate roughly 1.6 million annually from the switch. This would pay off what the authority plans to spend on the project within 8 years given the current bus replacement cycle.

VIA Metropolitan Transit in San Antonio has taken similar steps in converting its fleet of vehicles to natural gas consumption. In 2010, the transit authority introduced the first CNG vehicles to the San Antonio fleet. A year later, VIA was awarded a grant by the Federal Transit Administration to pursue the purchase of CNG vehicles to be used for the rapid transit service that would begin operation in late 2012. The VIA primo fleet consists of nineteen 60-foot CNG-powered buses.

Laredo reopened the city's compressed natural gas refueling station in the summer of 2011 using a federal grant provided through the State Energy Conservation Office. The city maintains a fleet of 63 compressed natural gas-powered vehicles, 32 of which are metro buses with the remainder being a mix of service vehicles. Additionally, the facility is open to public use for personal vehicles and private operations fleets (Diaz 2011).

LNG (Liquefied Natural Gas) Export

In addition to the increased use of domestic applications for natural gas that include power generation, manufacturing, and vehicle fuel, there is also the growing prospect for export. While this has remained a somewhat controversial topic, NERA Economic Consulting (which was commissioned by the Department of Energy) and Deloitte MarketPoint studies examining the impact of LNG export indicate that the effects would be a net plus for the US economy (NERA Economic Consulting 2012; Deloitte 2011), as well as Texas, and are unlikely to have a significant effect on domestic natural gas prices (Medlock 2012).

At present, global natural gas markets are not integrated. Prices vary from $0.75 per thousand cubic feet in Saudi Arabia to $4–6 in the United States to around $12 in Europe and as high as $16–17 in Japan. This situation is based on short-term shifts in supply and demand which have created export arbitrage opportunities (EIA 2012a).

In order to ship natural gas abroad from the United States efficiently, it must be supercooled to minus 260° Fahrenheit near an export terminal at a port and transformed into LNG, which reduces its volume by more than 600 times. An LNG tanker then transports the product to its designated foreign market. When the LNG reaches its destination, it is revaporized (or regasified) back into a gas before being shipped to its final destination by pipeline. Each step in this process is significant in terms of operating costs (Kawamoto 2008).

For example, given the current worldwide price differentials, it is profitable to ship LNG to Japan from the United States. Assuming a US market price of $4 per thousand cubic feet, there is the additional cost of approximately $6.40 to liquefy, transport, and regasify at the delivery point in Japan – more than doubling the price. Even so, a healthy profit of $6.60 for every thousand cubic feet is still generated (Henderson 2012). However, this lucrative opportunity will not go unnoticed by Australian, East African, and even Canadian natural gas suppliers – all of whom have substantial natural gas reserves and are equally or better positioned logistically to ship to Japan than is the United States

Similarly, prices in Europe have remained artificially high because of Russia's Gazprom monopoly on natural gas exports. With the threat of LNG imports from the United States, Ukraine, and other countries (Peaple 2013), prices in Europe are unlikely to remain at current levels either. Further, Gazprom's pipeline monopoly is already under siege from domestic producers in Russia, such as Novatek and Rosneft, and Statoil in Norway (Marson 2013).

In short, markets are dynamic. While there is an attractive export opportunity in the near term (3–5 years) for US producers, over the longer term, supply will likely catch up with demand and reduce price differentials (Medlock 2012). As such, first-mover advantages will accrue to those companies that can tap into the natural gas export market early.

The eventual synchronization of supply and demand would serve to both curb the demand for exports from the United States and put downward pressure on natural gas prices (Henderson 2012). In the same way that crude oil has become a global market, natural gas is likely to do so as well. This would come about as a direct result of new, significant natural gas discoveries and eventual production in the United States, Australia, East Africa, and probably China – perhaps other countries as well. These countries will seek to export their surplus or, in the case of China, reduce their need to import. Such developments imply increased global production, increased worldwide export, and greater price stability. As unconventional methods of natural gas production become dominant, the frequent shortages and price spikes – which occurred during the era of conventional exploration and production – will likely moderate.

Comments on Longevity of the Eagle Ford Shale Play

The activity in the Eagle Ford Shale grew from very little activity in 2008 to the significant levels described in this report for 2012 and for projections to 2022. Estimates for recoverable reserves in the region still vary widely. Oil estimates, for example, range from 3 to 10 billion barrels. Improvements in technology will play a key role in determining the final number. However, even on the basis of the low end of technically recoverable reserves, the estimates suggest that the Eagle Ford will continue to be a significant economic driver in South Texas for some time to come.

Certainly, any future activity in the Eagle Ford will be dependent on commodity prices. As mentioned previously in the report, we have seen the impact of low natural gas prices, which caused production to flatten out from 2011 to 2012. Any unexpected dip in oil prices could have a similar impact on crude oil production. However, over the long term, the prospect for continued activity in the Eagle Ford remains bright. As exploration and drilling activities give way to production and maintenance activities, the area is expected to settle into a "new normal" at some level higher than before production began.

The ultimate success of communities in the South Texas area will depend on how local leadership responds to the opportunities and challenges. In addition, area landowners, businesses, and communities will be recipients of significant amounts of unexpected wealth – how they steward this new wealth will have an important impact on the region's future. In the final analysis, economic development in the Eagle Ford should be balanced by quality of life, aesthetics, and environmental issues so that sustainable growth for the region is assured.

Acknowledgments The author wishes to thank Javier Oyakawa and Hisham Eid for their contributions to this chapter.

References

Albrecht D (2012) A comparison of metro and nonmetro incomes in a twenty-first century economy. J Rural Soc Sci 27(1):1–23

Atkinson R (2004) Reversing rural America's economic decline: the case for a national balanced growth strategy. Progressive Policy Institute

Baker T (2003) More ghost towns of Texas. University of Oklahoma Press, Norman

Barer D (2013) Amid opposition, South Texas coal mine approved. State Impact

Basich G (2011) Corpus Christi converting vehicles to natural gas. Government-fleet.com

Black T (2013) Buffett like Icahn reaping tank car boom from shale oil. Bloomberg

Casselman B, Gold R (2012) Cheap natural gas gives new hope to the rust belt. The Wall Street Journal. October 24, 2012

Christopherson S, Rightor N (2011) How should we think about the economic consequences of shale gas drilling? Cornell University working paper series: City and Regional Planning

Collette M (2013) Plastics manufacturer inks $1 billion contract for two Corpus Christi plants. Corpus Christi Caller-Times

Considine T, Watson R, Blumsack S (2011). The Pennsylvania Marcellus natural gas industry: status economic impacts and future potential, Pennsylvania State University, College of Earth and Mineral Sciences, Department of Energy and Mineral Engineering

Deller S, Chicoine S (1989) Economic diversification and the rural economy: evidence from consumer behavior. J Reg Anal Policy 19(2):41–55

Deloitte Market Point (2011) Made in America: the economic impact of LNG exports from the United States, 2011

DeWitt J (2012) Natural gas prices sent most fertilizer production overseas. Quad-City Times

Diaz F (2011) Natural gas station is reopened. The Laredo Sun

Dittrick P (2012) WoodMac: Eagle Ford 2013 spending to reach $28 billion. Oil Gas J 110(12)

Dukes RT (2012) Largest refineries in the U.S. benefit from Eagle Ford crude. EagleFordShale.com

NERA Economic Consulting (2012) Macroeconomic impacts of LNG exports from the United States. December 2012

Energy Information Administration (2012) Effect of increased natural gas exports on domestic energy markets. U.S. Department of Energy

Energy Information Administration (2012) Europe and Asia are the leading destinations for U.S. coal exports in 2012. U.S. Department of Energy

Garza V (2012) TCEQ offering $18M in natural gas vehicle grants. The Austin Business Journal

Gold R (2013) Shell CEO scripts a leading role for gas. The Wall Street Journal

Greenwood A (2012) Shale gas boosts US chems, trickles down to steel, tyres. ICIS.com

Henderson J (2012) The potential impact of North American LNG exports. The Oxford Institute for Energy Studies, from the Preface by Howard Rogers

Hiller J (2012) Eagle Ford counties facing a rough road toward repairs. FuelFix

Kawamoto H (2008) Natural gas regasification technologies. USCG Proceedings

Kinnaman T (2011) The economic impact of shale gas extraction: a review of existing studies. Ecol Econ 70(7):1243–1249

Klein B (2013) The economic impact of travel on Texas. Texas Tourism, Office of the Governor, Economic Development and Tourism. Dean Runyan and Associates, Portland

Marson J (2013) Gazprom warns of a drop in profit, driving down stock. The Wall Street Journal

Medlock K (2012) U.S. LNG exports: truth and consequence. James A. Baker III Institute for Public Policy, Rice University

Osgood J Jr, Opp S, Bernotsky R (2012) Yesterday's gains versus today's realities: lessons from 10 years of economic development practice. Econ Dev Q 26(4):334–350

Peaple A (2013) A new foreign policy for Gazprom. Heard on the Street. The Wall Street Journal

Platts (2012) The North American gas value chain: developments and opportunities

Portney K (2013) Taking sustainable cities seriously, 2nd edn. MIT Press, Cambridge

Reuters (2012) NRG Energy drops plan to build 800-MW coal-fired unit in Texas

Reuters (2013) U.S. coal-fired power plant retirements top 9,000 MW in 2012

Ross M (1999) The political economy of the resources curse. World Polit 51(2):297–322

Rumbach, A, 2011. Natural gas drilling in the Marcellus Shale: Potential impacts on the tourism economy of the Southern Tier. Prepared for the Southern Tier Central Regional Planning and Development Board.

Sachs J, Warner A (1995) Natural resource abundance and economic growth. National Bureau of Economic Research working paper no. 5398

Savage J (2012) Corpus Christi type A board approves $3M jobs incentive grant for Italian plastics plant. Corpus Christi Caller-Times

Smith M (2012a) Development spark catching in Gregory as TPCO plan construction begins. Corpus Christi Caller-Times

Smith R (2012b) "Coal-fired plants mothballed by gas glut. The Wall Street Journal

Smith R (2013) Can gas undo nuclear power? The Wall Street Journal

Stevens P (2003) Resource impact – Curse or blessing? Centre for Energy, Petroleum and Mineral Law and Policy, University of Dundee

Thompson J (2012) Booming shale gas production drive Texas petrochemical surge. Southwest Economy 4

Tunstall T, Oyakawa J, Eid H, Medina C, Green M Jr, Sanchchez I, Rivera R, Liru J, Morua D (2012) Economic Impact of the Eagle Ford Shale. University of Texas at San Antonio Institute for Economic Development

Tunstall T, Oyakawa J, Eid H, Abalos R, Wang T, Calderon E, Melara K (2013) Economic impact of the Eagle Ford Shale. University of Texas at San Antonio Institute for Economic Development

Vaughan V (2012a) CPS to buy natural gas plant. San Antonio Express News

Vaughan V (2012b) Valero to ship Eagle Ford crude to Quebec plant. San Antonio Express News

Weber J (2012) The effects of a natural gas boom on employment and income in Colorado, Texas and Wyoming. Energy Econ 34(5):1580–1588

Winters J (2011) Human capital and population growth in nonmetropolitan U.S. counties: the importance of college student migration. Econ Dev Q 25(4):353–365

Yergin D (2012) The real stimulus: low-cost natural gas. The Wall Street Journal

Communities Experiencing Shale Gas Development

Kathryn J. Brasier, Lisa Davis, Leland Glenna, Timothy W. Kelsey, Diane K. McLaughlin, Kai Schafft, Kristin Babbie, Catharine Biddle, Anne DeLessio-Parson, Danielle Rhubart, and Mark Suchyta

Abstract The influx of natural gas infrastructure and laborers associated with Marcellus Shale development has raised questions regarding the presence of a "boomtown" effect on Pennsylvania's rural communities. This chapter examines quantitative and qualitative data gathered from four Pennsylvania counties to assess how various social indicators, including housing, healthcare, education, crime, and residents' perceptions of their communities, have changed as a result of Marcellus Shale development. The quantitative data indicate that changes across many of these indicators are limited and difficult to distinguish from regional and long-term trends. Qualitative data resulting from focus groups, on the other hand, demonstrates substantial changes in how residents feel about their communities and their outlook for the future. The community implications of these findings are discussed, as well as the methodological challenges of studying rural communities experiencing unconventional development.

Introduction

The Marcellus Shale is a natural gas-bearing geological formation that lies beneath portions of Pennsylvania, New York, Ohio, Maryland, and West Virginia. Recent advances in hydraulic fracturing and horizontal drilling technologies have led to rapid expansion of the natural gas industry in Pennsylvania. Between 2005 and the end of 2013, 7,430 unconventional gas wells had been drilled in the Commonwealth (Pennsylvania Department of Environmental Protection 2014). This rapid development of the natural gas industry has generated excitement but also created significant concern about the

K.J. Brasier (✉) • L. Glenna • T.W. Kelsey • D.K. McLaughlin • K. Schafft •
K. Babbie • C. Biddle • A. DeLessio-Parson • D. Rhubart • M. Suchyta
Department of Agricultural Economics, Sociology and Education, The Pennsylvania University Park, State College, PA 16801, USA
e-mail: kbrasier@psu.edu; lad13@psu.edu; llg13@psu.edu; tkelsey@psu.edu; dkk@psu.edu; kas45@psu.edu; babbiekr@psu.edu; dcr185@psu.edu; suchytam@psu.edu

L. Davis
Pennsylvania Office of Rural Health, Department of Health Policy and Administration, The Pennsylvania State University, University Park, PA 16801, USA
e-mail: lad13@psu.edu

W.E. Hefley, Y. Wang (eds.), *Economics of Unconventional Shale Gas Development*, Natural Resource Management and Policy 45,
DOI 10.1007/978-3-319-11499-6_7

potential implications for the Commonwealth's communities, especially rural communities that can become overwhelmed by significant changes to the population.

This chapter describes research on community impacts of Marcellus Shale development in four Pennsylvania study counties (Bradford, Lycoming, Greene, Washington) with very high levels of natural gas extraction (see Brasier et al. 2014). The specific topics reported here include impacts on population, housing, health and healthcare access, K-12 education, and crime. We use a combination of publicly available data sources, mostly at the county level, and primary data collected via focus groups to characterize the changes within these four counties.

Natural Resource "Boomtowns"

Social scientists refer to rural communities experiencing rapid natural resource development as "boomtowns" (Brown et al. 2005; England and Albrecht 1984; Krannich 2012). As a natural resource extraction begins, the demand for labor grows. A portion of this labor is highly skilled, and the local labor force often cannot meet the demand. As a result, workers are brought into the region, while others are attracted to the region because of the new economic opportunities. As the population grows, local governmental services, infrastructure, and housing stock become stressed.

Prior research on the community impacts of "boomtowns" emphasizes the "social disruption" that can accompany the rapid population growth and change, often resulting in the magnification of social problems in the community, such as drug and alcohol abuse, domestic violence, mental health problems, and crime (Parkins and Angell 2011; Camasso and Wilkinson 1990; England and Albrecht 1984; Freudenburg and Jones 1991; Freudenburg et al. 1982; Kohrs and Dean 1974; Gilmore 1976). While sudden natural resource development can bring opportunities for many rural communities that are often economically stagnant, new populations also can place strains on housing availability, social and healthcare services, law enforcement, and schools.

Boomtown research also stresses that the risks and opportunities vary by community, by social position, and over time (Freudenburg 1984; Freudenburg and Wilson 2002; Gramling and Freudenburg 1990; Schafft et al. 2013). Youth may be particularly susceptible to negative perceptions of boomtown development (Freudenburg 1984; Seyfrit and Sadler-Hammer 1988), and long-term residents may find the social fabric of their communities changed or threatened by the sudden presence of "newcomers." Longitudinal boomtown research has suggested the need to recognize boom, bust, and recovery stages (Brown et al. 2005). Boomtown development based on nonrenewable energy resources faces the eventuality of a "bust" or a period when extraction declines due to resource depletion, technological change, or geopolitical trends that make extraction less economically attractive (Bunker and Ciccantel 2005; Freudenburg and Frickel 1994). Although a recovery stage has been documented in research by Brown et al. (2005), relatively little research exists to indicate the factors that allow communities to successfully adapt or whether the recovery leads to continued dependency on cycles of investment in resource extraction (Freudenburg 1992; Freudenburg and Wilson 2002; James and Aadland 2011; Rural Sociological Society 1993).

Boomtown research was largely developed by studying communities experiencing natural resource extraction (e.g., coal, uranium) and large-scale industrial developments (e.g., power plants) in the Intermountain West during the 1970s and 1980s. An important question guiding research on Marcellus Shale development is how applicable the boomtown model is to this region and how consistent the findings on community impacts will be with previous research (Brasier et al. 2011; Jacquet and Kaye 2014; Kinchy et al. 2014). Development in the northeast region of the United States is occurring over a larger and more diverse geographic region, within communities that are not as isolated as those in the Intermountain West. Some regions within Pennsylvania have a long history of oil and gas development, which has allowed for greater familiarity and a base from which economic activity can be built. It also has influenced the ability of particular landowners to benefit economically through leasing and royalty income, as the history of resource extraction in some areas has meant the severance of surface and subsurface rights. Further, the progression of the development of unconventional natural gas has differed in scope and pace from those earlier studies, affecting the scale and speed with which communities experience potential impacts and their ability to respond and adapt (Jacquet and Kay 2014). Although the boomtown research has provided an important model and has led Marcellus Shale researchers to focus on a consistent set of issues (i.e., population, housing, human services, crime), the direct application of the model needs greater scrutiny.

Community Impacts of Marcellus Shale Development

Researchers have begun to examine the community impacts of Marcellus Shale development on Pennsylvania and the surrounding states. As an emerging literature, the results are sometimes inconsistent and need continued verification as the field continues to grow and evolve. We provide a brief summary of this literature in relation to the topics reported here.

Housing

Changes to the housing market have been identified as the most critical, acute, and immediate impacts of boomtown development. Williamson and Kolb (2011) describe waves of workers with differing housing needs. The first wave of workers require temporary housing units (hotels, company-sponsored residential facilities, campgrounds, and rental housing), preferably those that provide housekeeping and meals. A second wave of workers, associated with company headquarters and regional offices, tends to occupy rental and owner-occupied units. The impacts within the larger housing market are significant, driving up the cost and limiting availability of all types of housing. As a result, families at the economic margins in these communities are pushed down the housing ladder toward units of lesser quality for higher prices or are forced out of the market altogether (Ooms et al. 2011; Williamson and Kolb 2011). Although most housing markets at the time of the study were affected, the most rural areas with high levels of activity were the most stressed, as they had relatively few affordable housing options prior to Marcellus Shale development and more barriers to quick development of new affordable housing options (Ooms et al. 2011; Williamson and Kolb 2011). These studies also documented additional consequences of a stressed housing market, including increased homelessness, difficulties for social service agencies that provide temporary housing, and challenges for child welfare agencies that assess living situations of children at risk (see also Brasier et al. 2011).

More recent quantitative research has shown mixed findings. Kelsey et al. (2012a, b) find that "Townships and boroughs with more Marcellus wells on average experienced larger average increases in market value than did those without Marcellus wells" (5).[1] In contrast, research by Farren et al. (2013) found that while fair market rent was higher in intensely drilled counties in Pennsylvania, there was no impact on median home value and vacancy rates. They further suggest that housing markets have recovered to a significant extent.

Health and Healthcare Services

Health impacts related to Marcellus Shale activity are of great concern to many, but there is a paucity of research in this area (Adgate et al. 2014; McDermott-Levy and Kaktins 2012). Several papers have outlined epidemiological mechanisms for tracing and identifying potential impacts on human health (Adgate et al. 2014; Steinzor et al. 2013). Others have used limited data sets to identify concerns and potential links to Marcellus Shale activity. Ferrar and colleagues (2013) documented self-reported health impacts and mental and physical health stressors perceived to result from Marcellus Shale development. The Southwest Pennsylvania Environmental Health Project found that the most common symptoms associated with drilling were skin rash or irritation, nausea or vomiting, abdominal pain, breathing difficulties or coughing, and nosebleeds (Ferrar et al. 2013). We could find no existing studies of impacts on healthcare services.

Education

A few recent studies have examined impacts of Marcellus Shale development on the provision of educational services, particularly K-12 education. Schafft et al. (2014a, b) using survey data supplemented by field-based interviews, and focus

[1] Assessed values are the taxable value of a property used to determine property taxes owed to counties, municipalities, and school districts.

groups found that while respondents in high drilling activity areas recognized benefits of economic expansion, they also recognized that this came at the cost of increased economic disparities, as well as pronounced strains on community infrastructure including housing and roads. Road damage and congestion interfered with school bus routes, while the bus drivers themselves were often tempted away from school district employment by trucking jobs in the industry. While respondents reported that the influx of workers placed strains on housing stock, very few schools experienced pronounced changes in enrollment size or student demographics given that most workers arriving from outside the area did not bring families.

Schafft et al. (2013) found that, in the context of what is often portrayed as a highly polarizing topic, the perception of shale development-related opportunity by high school administrators was directly correlated with the perception of social, environmental, and economic risk. They also found that the intensity of those perceptions – both risks and opportunities – corresponded with the level of drilling activity in their school districts.

Crime

To date, research on the effects of Marcellus Shale activity on crime has found mixed results. Kowalski and Zajac (2012) examined data on calls for service from the Pennsylvania State Police and arrest data from the Uniform Crime Reporting program through the Federal Bureau of Investigation. They reported no discernible longitudinal trends in their study counties (the seven counties with the highest number of wells drilled) or in comparison to other counties in the state. Food and Water Watch (2013) found that incidents of disorderly conduct increased in rural counties experiencing Marcellus Shale development at a higher rate than non-Marcellus counties.

Community Change

Brasier et al. (2011) found that the most prominent concern among leaders in communities experiencing natural gas development, particularly in the early stages, are worries about fundamental changes to the social relationships within the community and to the physical beauty of the places that residents call home (see also Perry 2012). Survey research also documents that the most highly rated components of communities are environmental quality, neighborliness, drinking water, and schools; these are the same areas community members see as potentially being negatively affected by Marcellus Shale development (Alter et al. 2010). The level of activity and the stage of development interact with key community characteristics – particularly population size, proximity to population centers, access to transportation networks, level of existing infrastructure, and extractive history – to influence the perceptions of the impacts across places (Brasier et al. 2011).

An important component of many rural communities in Pennsylvania and the Northeast region of the United States is the agrarian character of the landscape and economy. Agriculture may be particularly affected by Marcellus Shale development because farmers are owners of large tracts of land that are often targeted for energy extraction (Glenna et al. 2014). The increased leasing and royalty income can lead farmers to change their operations or stop farming altogether, which could have substantial impacts on the landscape (Brasier et al. 2011). This concern is supported by studies that indicate change in agriculture coincident with Marcellus Shale activity. Adams and Kelsey (2012) found an association between intensity of gas drilling and a decline in dairy cow numbers, suggesting the potential that dairy farming, a mainstay of some regions, may be undergoing rapid change in relation to natural gas development (see also Finkel et al. 2013). The Pennsylvania Center for Dairy Excellence found that natural gas leads to countervailing trends in agriculture, with some farmers reporting plans to modernize their dairy operations, others reporting plans to disinvest in dairy, and still others reporting plans to consider alternative forms of agriculture (Frey 2012). These findings suggest that dairy farming – and the agricultural landscape – could be changed substantially as a result of the influx of gas drilling revenues (Glenna et al. 2014). These changes to the landscape can have consequences for the ways in which residents think about and relate to their community.

The Marcellus Shale Impacts Study

Using the boomtown model as a framework, the research reported here describes community impacts across a range of features expected to change in communities experiencing rapid natural resource development, including population, housing, health and healthcare service provision, crime and the criminal justice system, and educational systems (see Brasier et al. 2014). We also discuss findings about impacts on a particular population – youth – because of concerns around out-migration of young people from rural communities. And we examine how perceptions of the community might be changing as a result of Marcellus Shale development.

Four counties, located in two regions of the state, are studied in this project – Bradford, Lycoming, Greene, and Washington. These counties have experienced some of the highest levels of Marcellus Shale development in Pennsylvania, yet they have diverse populations, histories, economic bases, and geographic locations. These differences allow comparisons that facilitate understanding of the potential effects of Marcellus Shale development across the Commonwealth. Regional comparisons also are made, by defining two regions as those counties that are adjacent to the four study counties. The Northern Tier contains 12 counties: the two study counties of Bradford, Lycoming, and ten neighboring counties (Clinton, Columbia, Montour, Northumberland, Potter, Sullivan, Susquehanna, Tioga, Union, and Wyoming). The southwest region consists of six counties: the two study counties of Greene and Washington and four neighboring counties (Allegheny, Beaver, Fayette, and Westmoreland).

The Four Study Counties

The four study counties were among the top six counties in the Commonwealth in the number of wells drilled from 2005 through the end of 2013 (Pennsylvania DEP 2014). Together, the four study counties account for nearly half (3,654) of the 7,430 unconventional wells drilled in the Commonwealth (Table 1). Washington County was the site of the first Marcellus Shale well, beginning production in 2005 (Harper 2008). Bradford County experienced significant growth starting in 2009 and had the highest level of activity until 2012, when drilling activity declined.

Table 2 offers an overview of the four study counties as of 2000, which provides important context for understanding differences between the counties and regions prior to Marcellus Shale development. Two counties – Lycoming and Washington – are classified as metropolitan by the US Census Bureau. Lycoming County is in the

Table 1 Study counties and wells drilled, 2005–2013 (Adapted from Brasier et al. 2014)

	2005	2006	2007	2008	2009	2010	2011	2012	2013	Total
Bradford	1	2	2	24	158	373	396	164	108	1,229
Washington	5	19	45	66	101	166	155	195	220	972
Lycoming	0	0	5	12	23	119	301	202	163	823
Greene	0	2	14	67	101	103	121	105	117	630
Total in study counties										3,654

Source: Pennsylvania Department of Environmental Protection, Office of Oil and Gas Management (Accessed May 21, 2014)

Table 2 Pre-Marcellus characteristics of study counties in 2000 (Adapted from Brasier et al. 2014)

	Population	People/sq. mi.	% employed in mining	% unemployed	Median household income (adjusted for 2012)
Northern Tier[a]	47,968	83	0.6	6.0	$47,071
Bradford	62,761	55	0.6	5.5	$48,451
Lycoming	120,044	97	0.1	6.3	$47,038
Southwest[a]	370,881	505	1.8	6.6	$47,901
Greene	40,672	71	6.7	9.2	$41,972
Washington	202,897	237	1.3	5.3	$52,004
Pennsylvania	12,281,054	274	0.3	5.7	$55,460

The Northern Tier region contains 12 counties: Bradford, Lycoming, and ten neighboring counties (Clinton, Columbia, Montour, Northumberland, Potter, Sullivan, Susquehanna, Tioga, Union, Wyoming). The Southwest region consists of six counties: Greene, Washington, and four neighboring counties (Allegheny, Beaver, Fayette, and Westmoreland)

Source: Social Explorer Tables 2011; Census 2000

[a]County average, includes study counties

Williamsport metropolitan area (2,000 population 120,044), and Washington County is part of the much larger Pittsburgh metropolitan area (2,000 population 2.4 million) (US Census 2000). In comparison, Bradford and Greene counties are classified by the USDA ERS as nonmetropolitan counties with small urban populations (2,500–19,999 people) adjacent to metropolitan areas (2013).

Methods of Data Collection and Analysis

This study uses a combination of quantitative and qualitative data to describe the impacts of Marcellus Shale activity within the four study counties. The quantitative analysis describes trends over time, comparing pre-Marcellus (prior to 2007) and early-Marcellus (2008–2012) data. In most cases, trends in the study counties are compared to trends in adjacent counties and for the state. The exception to this approach is for health and healthcare services, for which data were limited to the study counties only. Descriptive statistics/analyses are relied on as the primary form of analysis unless otherwise stated. In-depth, qualitative data, gathered primarily through focus groups, are used to interpret trends in the quantitative data. The use of multiple techniques gives breadth and depth to the case studies.

Quantitative Data Sources

Multiple data sources were acquired and analyzed for the study counties, as listed in Table 3. Most of these data were collected from publically available data sources via the Internet.[2] Data for each topic were analyzed at the county level, with the exception of the education data, which were organized by school district.

Qualitative Data Sources

Youth

Five focus groups were conducted with 36 youth in grade 11 in one school district from each of the study counties (one district set up two focus groups). This grade was targeted because this is a critical age in preparing for the transition from expectations to realization of careers and residential choice. School districts were selected through a multistage process. First, the number of wells in each school district was

[2]The exceptions to this were calls-for-service data (acquired directly from the Pennsylvania State Police) and the Emergency Medical Services data (provided by the Pennsylvania Trauma Systems Foundation under an agreement with the Pennsylvania Department of Health's Bureau of Emergency Medical Services).

Table 3 Data sources used by topic

Topic	Data sources
Population	2000 Census, Social Explorer Tables: US Census Bureau
	American Community Survey 2005/2007 (3-Year Estimates)
	US Census Bureau
	2010 Census: Social Explorer Tables: US Census Bureau
Housing	2000 Census, Social Explorer Tables: US Census Bureau
	American Community Survey 2005/2007 (3-Year Estimates)
	US Census Bureau
	2010 Census: Social Explorer Tables: US Census Bureau
	American Community Survey 2009/2011 (3-Year Estimates)
	US Census Bureau
Health and healthcare utilization	Health Care Cost Containment Council
	US Department of Health and Human Services, Centers for Medicare and Medicaid Services
	US Department of Health and Human Services, Office of Women's Health: Quick Health Data Online
	Pennsylvania Department Public Welfare
	Pennsylvania Trauma Systems Foundation, Pennsylvania Department of Health, Bureau of Emergency Medical Services
Education	Pennsylvania State Department of Education
	National Center for Education Statistics, Common Core of Data
Crime	Pennsylvania State Police Calls for Service
	Federal Bureau of Investigation, US Department of Justice. Uniform Crime Reporting Program Data
	Common Pleas Case Management System
	Magisterial Court Data

calculated (Pennsylvania Department of Environmental Protection Office of Oil and Gas Management 2014). The three districts in each county with the highest well count were identified; then one district per county was randomly selected. Administrators in each of the four districts were asked to identify a group of between six and eight 11th graders, with an even gender representation, representing more or less demographically "typical" students within the district.[3]

School Districts

In the same school districts selected for the youth focus groups, we also held one focus group with educators (one district arranged two educator focus groups). These five focus groups were supplemented by two focus groups with school

[3] In all but one focus group, no school administrators were present. In the focus group in which administrators were present, students were notably more positive in their assessments of Marcellus community impacts. It is not possible to estimate the influence of the administrators' presence.

district superintendents, one in each region. In addition, we held one focus group with vocational educators. In total, we held eight focus groups with 47 educators and administrators.

Community Leaders and Human Service Agency Representatives

A total of eight focus groups including 36 participants were conducted between November of 2012 and June of 2013. In each region, four focus groups were conducted that included representatives of local government; health, housing, and human service agencies; local economic development agencies and businesses; and farmers and representatives of local agricultural businesses and service agencies. Potential participants for focus groups were identified through key informants such as Cooperative Extension educators and agency representatives, recommendations and referrals from elected officials and professional contacts, and focused internet searches.

All of the focus groups were audio recorded; the recordings were then transcribed. Using a constant comparative method (see Corbin and Strauss 2008; Creswell 2013), transcriptions and field notes were open coded for perceptions of community change. Particular attention was paid to changes in categories that reflected our quantitative analyses, including population change, economic conditions, and pressures on social services. The research team developed the coding scheme collaboratively, reconciling initial differences in interpretation through discussion and consensus building.

Results

The following section summarizes the findings using both the quantitative and qualitative data for all four counties.

Population

The boomtown model predicts that natural resource development would lead to substantial population growth. For Marcellus Shale development, this growth may be driven by an influx of workers in the natural gas and ancillary industries, as well as the arrival of family members of these workers and other individuals attracted to expanding economic opportunities. Documenting population change is critical, as demographic shifts influence demand for services, economic activities, infrastructure, and housing. In Table 4, we examine population changes for the four study counties using county-level aggregate data and estimates from the US Census of Population and Housing (2000 and 2010) and the American Community Survey (3-year estimates, 2005–2007).

Table 4 Population change in Pennsylvania, study counties, and region, 2000–2010 (Adapted from McLaughlin et al. 2014b)

	Population (avg. annual rate of change[a])		
	2000–2005/2007	2005/2007–2010	2000–2010
Pennsylvania	12,281,054 (1.6)	12,400,959 (6.1)	12,702,379 (3.4)
Northern Tier[b]	532,741 (−2.3)	525,508 (6.1)	538,354 (1.1)
Bradford	62,761 (−3.0)	61,626 (4.0)	62,622 (−0.2)
Lycoming	120,044 (−3.8)	117,311 (−2.6)	116,111 (−3.3)
Southwest[c]	2,225,284 (−5.4)	2,153,833 (−1.4)	2,142,168 (−3.7)
Greene	40,672 (−3.9)	39,717 (−6.5)	38,686 (−4.9)
Washington	202,897 (2.0)	205,302 (3.1)	207,820 (2.4)

Sources: US Census Bureau, 2000 Census, 2005–2007 ACS 3-Year Estimates, 2010 Census

[a]Measured as average change in number of residents per 1,000 for each year in period. Midpoint of 3-year estimates, 2006, is used to determine number of years in period

[b]Northeast Marcellus region includes 12 counties: the two study counties (Bradford and Lycoming), adjacent counties (Clinton, Columbia, Northumberland, Susquehanna, Tioga, Union, Wyoming), and three counties (Montour, Potter, and Sullivan) that were excluded in this analysis because their population counts were too small to be estimated in the 2005/2007 ACS data

[c]Southwest region includes six counties: two study counties (Greene and Washington) and four adjacent counties (Allegheny, Beaver, Fayette, and Westmoreland)

Bradford County experienced significant population growth coincident with the timing of Marcellus development. Bradford County lost over 1,000 residents from 2000 to 2005/2007 but regained nearly the same number from 2005/2007 to 2010. In contrast, Lycoming County's population shrank from 2000 to 2010. Lycoming County's loss of nearly 4,000 residents over the decade is particularly striking given the surrounding regional and state contexts, although the rate of loss was less pronounced from 2005/2007 to 2010 (−2.6) than earlier in the decade (−3.8).

Greene County's population declined throughout the decade with the rate of decline actually accelerating in the latter half of the decade, from a rate of 3.9 residents per 1,000 population per year from 2000 to 2005/2007 to a rate of 6.5 persons per 1,000 population per year from 2005/2007 to 2010. While the county lost only 2,000 residents across the decade, this is a relatively large share – about 5 % – of its small population. In sharp contrast, Washington County added nearly 5,000 residents over the decade, with the rate increasing slightly in the latter half of the decade, coincident with Marcellus Shale development.

Of the four study counties, Bradford and Washington Counties experienced population growth coincident with Marcellus development. The growth was most pronounced in Bradford County, where Marcellus activity may have contributed to a turnaround from a loss of population early in the decade (2000–2005/2007) to a gain in population later in the decade (2005/2007–2010). Among our study counties, Bradford had the most wells drilled by the end of 2010, with more than 550 wells drilled from 2005 to 2010. Washington County's population growth is similar to the state trend and may be a continuation of the pattern since 2000. The analysis suggests that long-term trends in population growth or loss are not easily changed by the introduction of one economic activity.

This analysis of population change has some important limitations, most particularly related to the data. The Census and the American Community Survey are derived from surveys of households conducted at a particular point in time. Those living in temporary housing (e.g., RV parks, campgrounds, hotels), which is typical of some proportion of workers in this industry, will not be counted as residents. The county-level trends shown here also do not indicate *how* the increases or decreases occur, as the population counts are a result of a combination of in-migration, out-migration, and retention of current residents. Relatedly, it is unclear how population flows related to Marcellus Shale development interact with other trends affecting population change. These trends include in-migration of residents from New York and New Jersey to northeastern Pennsylvania, long-term trends of population loss in large urban areas and western Pennsylvania, and the location of new large-scale facilities (e.g., casinos) in some regions.

Housing

The boomtown model would suggest that the housing market becomes stressed because of the immediate need to house the influx of workers associated with the industry. The largest housing impacts of Marcellus activity are likely to be seen in the counties with the most active drilling and the smallest populations – Bradford and Greene in this study. In contrast, larger population counties are believed to have more capacity to absorb new residents because they have a larger housing stock and more capacity to build new housing. In this study, Lycoming and Washington counties fall in this category. The focus group data bear this out. Housing was identified by one Northern Tier focus group participant "*the biggest issue.*" In the Northern Tier focus group, one participant indicated that their organization conducted a survey of residents every 3 years, *the housing issue, in 2006 it was something like 21st or 20th [among respondent concerns] and then it rose to 7th in 2009, and it's first in 2012.* The location of the activity also matters, as regional headquarters and staging areas can be located proximate to (but not necessarily in) areas where drilling is occurring and where there is adequate access to needed infrastructure (transportation, commercial space, worker housing). Both Lycoming (Williamsport) and Washington (Canonsburg) counties are home to headquarters of gas companies that serve the surrounding region. This dispersion of gas-related activity to counties other than those experiencing the largest volume of active drilling may result in smaller differences observed between counties with active drilling and those with minimal or no drilling activity.

We analyze changes in housing stock, vacancy rates, and housing affordability in the four study counties within the regional and statewide contexts. Data from the US Census of Population and Housing (2000) and the American Community Survey (ACS 3-year estimates, 2005/2007 and 2009/2011) were used for these analyses. We also draw on the focus group data to describe the perceptions of impacts on housing that are not readily seen in the quantitative data.

Housing Stock

The number of housing units in Bradford County increased 2.1 % from 2000 through 2005/2007 and 2.5 % between 2005/2007 and 2009/2011. The percentage increase in the beginning period, pre-Marcellus, is below that of the region (2.8 %) and the state (3.8 %); the increase in the latter period, during Marcellus development, is higher than either the state (2.2 %) or the region (2.0 %). Lycoming County experienced a growth of 2.4 % in the first half of the decade followed by a 2.4 % decrease in housing units in the latter part of the decade, coincident with Marcellus Shale development; both of these percentages are lower than the state and the region change in housing stock.

Greene County experienced an increase of 2.7 % housing in the early part of the decade (2000–2005/2007) and a loss of 4.0 % in the latter part of the decade (2005/2007–2009/2011). The earlier increase is higher than that of the region (1.9 %) but lower than the state. The decrease in the latter part of the decade is larger than that of the region (−0.4 %) and stands in contrast to the growth in the rest of the state (2.2 %). Washington County had an increase in housing units of 4.9 %, the highest of the study counties (2000–2005/2007); this was followed by an increase of 1.6 % (2005/2007–2009/11). Both changes are higher than the region.

The lower growth changes in housing stock in the latter part of the decade are heavily influenced by the housing crisis and recession. In some communities, a lack of affordable housing before Marcellus Shale development was exacerbated by the influx of gas-related workers (Lycoming County Department of Planning and Community Development 2012). In the southwest, the influx of gas workers was preceded by foreclosures that further limited the availability of affordable housing. *The people that were foreclosed upon took the housing that the low income could afford, because they had nowhere else to go*. Because of this, it becomes difficult to assess the influence of Marcellus Shale development. However, Bradford County's increase of 2.5 % in housing stock, at a time when the region and state were growing at lower rates, suggests the potential that housing stock was increasing in reaction to increased demand related to natural gas development.

Occupancy and Vacancy of Housing Units

An increased demand for housing would likely be reflected in an increase in owner-occupied and rented units and a decrease in the vacancy rate. In Bradford County, owner-occupied housing declined by 3.2 percentage points, with most of that occurring between 2005/2007 and 2009/2011. The percentage of housing rented increased from 20.9 % in 2000 to 21.6 % in 2005/2007 then declined to 18.9 % in 2009/2011. In contrast, the vacancy rate increased in each time period, with the largest increase (4.8 % points) occurring from 2005/2007 to 2009/2011. Lycoming County also experienced a decline in the percentage of owner-occupied units from 2000 (62.2 %) to 2005/2007 (60.5 %), then stayed relatively steady at 60.8 % in 2009/2011. The percentage of rented units in Lycoming County, like Bradford County, increased in the early part of the decade (0.9 %, from 27.4 to 28.3 %) then declined between 2005/2007 and 2009/2011 (to 27.6 %). The percent of vacant units increased across the decade, from 10.4 to 11.2 to 11.6 %. The overall trends for both Bradford and Lycoming County, although similar to the Northern Tier, suggests either a decline in demand for rental or owner-occupied units, an increase in supply (as noted above), or that housing has been priced beyond the ability of people to pay.

Owner-occupied housing in Greene County declined from 66.9 % in 2000 to 63.1 % in 2009/2011. The percentage of housing units rented dropped from 23.4 % in 2000 to 19.5 % in 2005/2007, but then increased to 24.2 % in 2009/2011. This suggests volatility in the demand and/or availability of rental units in the county. Vacant housing in Greene County rose markedly from 9.7 % to 15.3 % from 2000 to 2005/2007 and then dropped to 12.7 % by 2009/2011. During this time, the housing stock in Greene County rose and then declined to below that available in 2000.

Although the general patterns for the rates of owner-occupied and rented housing in Washington County are similar to those in Greene County, there is greater stability overall. The percentage of owner-occupied housing decreased from 2000 (71.7 %) to 2005/2007 (70.8 %) to 2009/2011 (69.6 %); the percent rental decreased from 21.3 to 19.8 % then increased to 20.7 %. The percentage of housing vacant increased in Washington County from 7.0 % in 2000 to 9.7 % in 2009/2011. Washington County had relative stability in housing occupancy and vacancy over the period studied, perhaps partly because of the much larger size of the housing stock and the increase in housing of almost 6,000 units over this time frame. The variations in vacancy of housing in Bradford and Washington counties may reflect differences in the timing and/or types of Marcellus-related activity in these two counties. Or they may reflect other development activities, suburban expansion in Washington County, or differences in the demolition of dilapidated housing.

Housing Affordability

Increased demand for housing can lead to higher rental rates and decreased housing affordability. Housing is considered affordable when a household spends no more than 30 % of its annual income on rent (United States Department of Housing and Urban Development 2014). Across Pennsylvania, the percentage of households spending 30 % or more of annual income on rent increased by 10.5 % points from 2000 to 2009/2011, to 46.1 % at the end of that period.

All four study counties experienced an increase in the percentage of households spending more than 30 % of their income on rent between 2000 and 2009/2011 (Table 5). Bradford County had 29 % of renter households lived in housing that was not affordable in 2000 and 36.7 % in 2009/2011. Lycoming County experienced a larger increase (10.6 % age points) in the percentage of households living in unaffordable housing, from 35.6 % in 2000 to 46.2 % in 2009/2011. Greene County experienced the largest increase, from 32.5 % in 2000 to 46.1 % in 2009/2011. Washington County experienced a 9.3 % point increase in the percent of households paying above 30 % of their income in rent, from 33.5 % in 2000 to 42.8 % in

Table 5 Percentage of renters spending more than 30 % of income on rent (Adapted from McLaughlin et al. 2014a)

	2000	2005/2007	2009/2011	% change 2000–2005/2007	% change 2005/2007–2009/2011
Pennsylvania	35.6	43.1	46.1	7.5	3.0
Northern Tier[a]	32.5	36.8	41.6	4.3	4.8
Bradford	29.0	36.2	36.7	7.2	.5
Lycoming	35.6	43.9	46.2	8.3	2.3
Southwest[b]	33.3	40.1	42.1	6.8	2.0
Greene	32.5	41.5	46.1	9.0	0.6
Washington	33.5	39.3	42.8	5.8	3.5

Source: Social Explorer Tables. Census 2000, ACS 2005 to 2007 and ACS 2009 to 2011 (3-Year Estimates), Social Explorer; US Census Bureau

[a]County average for nine counties in region. Montour, Potter, and Sullivan counties also were adjacent to Bradford and Lycoming counties but are excluded because their population counts were too small to be estimated in the 2005/2007 ACS data

[b]County average for six-county region (includes study counties)

Table 6 Median household income by tenure (adjusted to 2012 dollars), study counties, regions, and the state (Adapted from McLaughlin et al. 2014a)

	2000	2005/2007	2009/2011	2000	2005/2007	2009/2011
	Owners	Owners	Owners	Renters	Renters	Renters
Pennsylvania	$65,838	$65,003	$64,861	$34,019	$29,830	$28,928
Northern Tier[a]	$54,754	$52,643	$53,153	$28,460	$26,395	$25,751
Bradford	$54,836	$48,760	$52,548	$30,407	$27,174	$25,689
Lycoming	$56,599	$55,405	$54,607	$29,521	$25,495	$24,614
Southwest[a]	$56,930	$57,181	$57,915	$27,624	$24,377	$24,332
Greene	$50,422	$49,251	$54,487	$22,406	$19,379	$22,177
Washington	$60,604	$63,398	$62,371	$28,282	$24,990	$24,820

Source: Social Explorer Tables. Census 2000, ACS 2005 to 2007 and ACS 2009 to 2011 (3-Year Estimates), Social Explorer; US Census Bureau

[a]County average for the region, includes study counties

2009/2011. The percentage change is largest from 2000 to 2005/2007 for all four study counties, the state, and the southwest region; the Northern Tier experienced a larger increase in the latter time period.

These changes in affordable housing need to be understood in the context of changing household incomes in the study counties (Table 6). In Bradford County, median household income dipped from 2000 to 2005/2007 before rising again. In Lycoming County, median household incomes of owners declined slightly across each time point. Greene County median household income of owners declined slightly from 2000 to 2005–2007, but then rose in 2009–2011. Washington County's owner-occupied households experienced a rise in median household income from 2000 to 2005–2007, but median income then declined slightly by 2009–2011.

Renters in Bradford, Lycoming, and Washington counties fared poorly over this time period, with lower inflation-adjusted median household income in 2009/2011 than in 2000. Only Greene County renters had median household incomes in 2009/2011 that were comparable to those in 2000. The information on median household income for those in owner-occupied units and those in rental units suggests that the decreasing affordability of rental housing can be attributed, in part, to lower inflation-adjusted incomes among renters.

Focus group participants suggest that these figures likely understate the extent of volatility in the housing market, declining household incomes, and declining housing affordability. The focus group participants described how demand for housing by gas workers has led to increased rental rates, rates that local residents are unable to afford. One focus group participant in the southwest stated,

> Before Marcellus Shale, we didn't have many problems with clients finding homes. The landlords worked with the low income….prior to Marcellus Shale, you could probably find an apartment for $300 or $400 easily….but now it's … in some cases impossible for people to find housing.

Industry workers often receive a housing allowance and earn higher incomes than residents, allowing them to pay higher rents. A southwest focus group participant reported …*The people that are able to afford the $1,400 a month rent is because they might be lucky enough to be in the one job….that's offering them the prevailing wage….where they're getting $150 a day for living expenses.* A Northern Tier participant remarked upon the increase in rents noting, *We're seeing…the evictions… with the housing costs going up… and people aren't able to afford it, so families are moving out…* Another participant recounted how a landlord in a mobile home park evicted all current residents, improved the homes, then rented to gas workers at higher rates. The displaced residents lost their homes and community.

Focus group participants described how landlords can charge gas-related workers more than they can recoup from low-income families on housing assistance. Focus group participants in both regions commented on the waiting lists for housing that will accept vouchers: *We have seen our public housing waiting list almost triple since…2008.* Low-income families move in with other family members, living in tents or cars, or leave the area in search of affordable housing. As one Northern Tier participant described: *You're not seeing homeless people sleeping on the street, but they're couch surfing. They are living with their parents. They are living multiple households in one structure…They're just finding shelter wherever they can… we're talking about whole families….being displaced.* Participants described increased demand – and related stress – on homeless shelters, including one in the Northern Tier that had sheltered 450 individuals, "…*100 of them … children.*"

The lack of quality housing stock in some areas contributes to problems for low-income families. One focus group participant in the southwest observed that:

> Once they've (the gas workers) moved in, they've basically destroyed the apartments. Then, the landlords don't want to redo the apartments, and they've raised the rent from…maybe $400 to $1,000 a month. Now the landlord wants that money again, and …your low income….there's no way they can afford that.

Focus group participants suggest that overall, housing is less available, more expensive, and of lower quality. Low-income individuals and families have been displaced with few options but to find other temporary housing situations locally, move elsewhere, or become homeless.

The displacement of low-income families and individuals noted by the focus group results suggest that the quantitative data need to be viewed provisionally. The Census and ACS data do not reflect the experiences of those who reside in temporary housing or who have been displaced because of increases in rent or loss of income.

Health and Healthcare Utilization

It is difficult to definitely ascertain that changes in health status or healthcare utilization are a direct result of Marcellus Shale drilling activity. The primary issue here is that the healthcare delivery system, as a general rule, does not collect data from patients on their employment status or whether they are employed in an occupation associated with drilling. If that type of information is collected, it is not done on a systematic basis and is not publically available. The same holds true with human and social service agencies but perhaps to a lesser extent. The providers may, due to the more personal nature of the services they provide, have greater knowledge of the employment status of their clients, but, like healthcare delivery systems, they do not collect, or report, that information in a consistent, quantifiable, publically available manner. As a result of these limitations, the results are based largely on the association between the data and the time frame of Marcellus drilling activity. Here, data on healthcare service utilization, insurance status, injuries by type, and emergency medical service complaints are examined and compared during the period prior to and during Marcellus Shale drilling activity expansion to identify any potential associations.

Access to Healthcare Services

Access to comprehensive, quality healthcare services impacts an individual's physical, social, and mental health status, contributes to preventing disease, assists in detecting health conditions, facilitates treatment for health issues, and improves quality of life and life expectancy. As such, the analysis focuses on the use of and access to healthcare providers, including hospital services, community-based (safety net) care, emergency care, and sources of insurance through Medicare and Medicaid as well as percentages of the population who are uninsured (Centers for Medicare and Medicaid Services 2013a, b, c).

The study counties saw slight variations in the number of acute and community-based healthcare services. The number of hospitals remained steady, with two in the southwest and seven in the Northern Tier. Two in the Northern Tier are federally designated Critical Access Hospitals meaning that they must be in a designated rural area, have 25 beds or less, and meet other criteria. Inpatient hospitalizations in the four counties and the two regions increased slightly in the Northern Tier and decreased slightly in the southwest, but it is not possible to directly connect this to Marcellus Shale drilling Davis et al. 2014).

The number of "safety net providers" increased slightly in the study years, by four federally qualified health centers (FQHCs), by the establishment of a Rural Health Clinic (RHC) in one county, and by the closure of an RHC in another county (Davis et al. 2014). The numbers of these providers does not seem to be associated with a change in population overall but may reflect an increase in the uninsured population in certain counties and federally funding available to support the establishment or expansion of these types of community-based healthcare providers. Access to primary care providers was and continues to remain an issue, and the demand for mental and behavioral health services has increased as have the interagency strategies for addressing this need. Input obtained from focus groups offer additional insight into the need for and use of healthcare services for primary and emergency care services and the lack of some services and strategies the agencies have used to respond to increased demands. Noted one participant, *[F]or a while you couldn't get a dentist appointment within an hour's drive.* Focus group participants also addressed the need for mental and behavioral health services which they perceive to be an increased burden on county human services.

Injuries and Emergency Services

Injuries affect the population by imposing individual, social, and economic costs on society (Boden et al. 2001). How and why injuries occur provide important information for policymakers to design and focus intervention efforts to prevent injuries (Centers for Disease Control and Prevention National Center for Health Statistics 2012). For this analysis, 12 different types of injuries were examined during the period of 2000–2011: motor vehicle accidents, motorcycles, pedestrians, gunshot wounds, stabs, falls, hot/corrosive materials, fire/flames, struck by, caught between, machinery/power tools, and assaults. These are organized into categories that relate to increases in activities that might be associated with Marcellus Shale development. For example, an increase in traffic and pedestrian accidents could occur because of the increase in traffic, especially truck traffic, associated with drilling and pipeline construction. Injuries associated with crime (gunshot wounds and stabbings) might occur with an increase in the number of younger men in the population. Finally, the types of injuries associated with mechanisms such as being struck by, caught between, and machinery or power tools also might be more frequent at workplaces such as drilling rigs and other types of large machinery.

There are no overall trends for injuries in the four study counties; however, there are noticeable increases in injuries associated with falls, motor vehicle accidents, and accidents involving motorcycles. These types of injuries could be related to any type of large-scale construction activity such as Marcellus Shale development

(Davis et al. 2014). The increase in accidents could be associated with the increased vehicle traffic that results from an increase in population (and is supported by Food and Water Watch 2013). This conclusion was reflected in comments made by several focus group participants. One professional from the southwest region reflected, *In our county…we've had a huge increase in DUI charges. Last year alone, we had 25 individuals from out of state that were pulled over for a DUI.*

Emergency Medical Services Complaints

Emergency Medical Services (2013) is a system of coordinated response and emergency medical care involving several individuals and agencies. EMS is activated by an incident that causes serious illness or injury. Analyzing EMS complaints (the type of health issue an individual was experiencing when the EMS service was requested) can identify the types of health emergencies individuals were experiencing over time.

All four counties experienced substantial increases in the number of complaints, in some cases increasing by more than 3,000 %. For example, the number of complaints increased from 4,464 in 2009 to 11,819 in 2011 in Lycoming County; Bradford County increased from 1,646 in 2009 to 8607 in 2010; Washington County increased from 2,732 in 2009 to 33,632 in 2011; and Greene County increased from 149 in 2009 to 5,030 in 2011 (Emergency Medical Services 2013). Focus group participants did not address reported injuries per se and instead emphasized issues associated with individual behavior, such as drug use, outcomes associated with alcohol consumption such as DUIs, and sexually transmitted diseases. It is not possible to associate these comments with the data for complaints, but it can be hypothesized that behaviors associated with drug use could lead to these types of reported injuries.

> I'll tell you, one of the biggest problems that I know of in our business because we deal with some of these oil and gas people is this dope. They can't pass a drug test. They'll have an application out; there will be 25 people coming in, make an application out. Twenty-three of them can't pass the drug test.

Education

For schools based in communities experiencing boomtown-related growth, some of the most pressing questions involve effects on enrollments and student demographics, academic outcomes, and fiscal conditions at the school and district level. To some extent, these are all related issues. For example, will influxes of workers include families and children, and if so will this result in enrollment spikes? As a Lycoming County educator remarked in a focus group, initially "the enrollment numbers from PDE [the Pennsylvania Department of Education] were coming that we were gonna have 20 %, 25 % increase in students." In many rural school districts in Pennsylvania's

Marcellus region, gradual enrollment declines have been the norm over the last few decades and educators and administrators early in the process of shale gas development saw the potential for enrollment increases as a clearly beneficial outcome that might help to stave off school closures and consolidations that had been seen as inevitable. At the same time, there were concerns about how schools might be able to absorb sudden influxes of students and what kinds of needs new students might have. Others questioned how these changes might affect school district budgets, and hiring needs, and to what extent might the broader economic benefit accrue to local school districts.

School Enrollments and Student Demographics

To the surprise of educators, there were few changes in enrollment, student demographics, and student outcomes. Enrollments continued to exhibit steady and slow enrollment declines. Enrollment declined from the 2005–2006 to the 2010–2011 school years: 7.8 % in Bradford County, 3.8 % in Lycoming County, 8.0 % in Greene County, and 1.8 % in Washington County. The educators we spoke with consistently attributed this to out-of-state gas workers arriving without family members or children. In the Northern Tier focus group, one educator said:

> We probably have more wells than most (but) I hardly saw any kids. I mean, if I saw more than seven or eight kids over this whole time period that I could look at and say, "This is from the industry," we did well. Yet my enrollment continued to go down. The guys did not bring their families is what happened. They came. Their families stayed in Oklahoma or Texas or Louisiana. Then to top that off, there was no place in our school district for them to stay anyway.

While educators mentioned the presence of some new students as a consequence of incoming gas workers, the overall numbers of new students in nearly every district were low. Relatively steady or even declining total enrollments can hide student turnover, however, which could still affect the demographic makeup of student populations. However, county-level data does not suggest marked student demographic change or changes in English Language Learner (ELL) student populations or the percentages of students classified as receiving special education services (Schafft et al. 2014b).

Student Economic Status

During the second half of the 2000s, the statewide percentages of students' income qualified for free or reduced price lunch increased markedly, 23.8 % between the 2005–2006 and 2010–2011 school years. This increase is attributable in large part to the national recession and economic downturn. These rates rose within the study counties as well. The increase was below the state in three of the four study counties (5.3 % in Bradford County, 8.1 % in Washington County, and 10.6 % in Lycoming County) and above the state (26.6 %) in Greene County.

Academic Outcomes

A concern expressed by several educators was the extent to which lucrative employment in the industry might lead to increased dropout rates among students. In Bradford County, an educator said that the industry, for example, is *always looking for welders. So these kids, if they can go get certified to weld, they're gonna make more money than you and I are.* Another explained:

> There was like this mentality that, "Oh, I can make a lot of money. This is gonna be here forever. I've got it made. I'm 17, 18 years old. I could even quit school." We had a lot of fear that that was gonna have an impact on our graduation percentages, our percentage of kids going on to college and that kind of thing."

Youth as well were aware of the potential opportunities associated with the industry. A Bradford youth said, *I've already had an offer when I turn 18. I can start out making $3,000 a week from just as an assistant, and you don't even need a degree for that.*

However, dropout rates in the four study counties changed little from 2007–2008 to 2011–2012. The rates changed from 1.6 to 2.6 in Greene County, 1.1–1.4 in Washington County, 2.2–2.4 in Lycoming County, and 2.4 to 2.1 in Bradford County. Anecdotal evidence from focus groups suggests that some students have dropped out, lured at least in part by industry opportunities. An important question remains regarding how industry-associated opportunities may have affected postsecondary educational aspirations and how these aspirations may change over time with changes in the structure of the natural gas industry in Pennsylvania. Our data did not allow us to investigate these questions, but these trends and the effects on communities, schools, and other local institutions bear continued attention.

Crime

The impacts of Marcellus Shale activity on crime potentially could be felt in multiple ways, including criminal activity, criminal investigations, prosecution, and incarceration. The quantitative data examined here include calls for service handled by the Pennsylvania State Police, arrests by type of violations, new criminal and civil cases filed in judicial system, and traffic violations.

Pennsylvania State Police Calls for Service

There are differences in the levels and types of coverage provided by the Pennsylvania State Police across municipalities, suggesting that county-to-county comparisons are difficult to interpret. However, a comparison of each county's annual average rates of incidents before (2001–2007) and during (2008–2010) Marcellus Shale development for which Pennsylvania State Police responded indicates a rise in these rates for two study counties, Bradford and Washington (Table 7). The average annual rate in

Table 7 Average annual rates (per 1,000 Residents) of crime indicators for study counties prior to Marcellus (2001–2007, top number) and during Marcellus (2008–2010, bottom number) activity (Adapted from Brasier and Rhubart 2014)

	Bradford	Lycoming	Northern Tier*	Greene	Wash.	Southwest Region*	All PA Counties
PA state police	75.4	71.0	106.5	139.2	63.4	74.5	48.1
Calls for service	88.9	69.0	103.5	130.9	71.7	74.0	47.3
Arrests for serious	5.8	8.9	7.2	7.2	8.1	9.2	9.9
crimes	5.7	8.9	6.9	5.5	7.5	9.0	9.8
Arrests for minor	19.0	26.8	22.6	25.5	21.9	23.1	22.7
offenses	19.5	24.2	20.2	19.8	16.7	20.5	20.4
Arrests for driving	3.4	4.5	3.7	6.0	4.1	3.8	3.9
Under the influence	3.7	5.1	4.3	5.9	3.9	4.1	4.2
Arrests for drug	1.6	2.1	2.7	2.5	2.1	2.4	1.7
Abuse violations	2.0	1.7	2.9	2.4	1.6	2.5	1.5
New criminal	9.5	16.5	11.2	13.1	12.7	13.3	12.5
Cases filed	10.3	16.3	12.5	13.4	13.8	14.3	13.2
New civil cases	6.5	7.2	5.5	5.0	1.8	6.6	5.9
Filed	10.3	11.4	9.4	9.6	4.3	11.0	10.7
Traffic violations	89.1	105.5	151.3	130.9	139.3	124.6	153.0
	100.1	112.9	155.6	140.8	179.9	148.4	157.0

*=County average, includes study counties

Bradford County is 75.4 arrests per 1,000 residents 2001–2007 compared to 88.9 arrests per 1,000 residents 2008–2010. The average annual rate in Washington County is 63.4 arrests per 1,000 residents 2001–2007 compared to 71.7 arrests per 1,000 residents 2008–2010. Both Lycoming County and Greene County had average annual rates that were lower during the years of active well development than the predrilling period.

Arrests by Type of Crime

The arrest rates by type of crime across the study counties show a mixed and inconclusive picture of impacts of Marcellus Shale development on criminal activity. The arrest rates in Bradford County do not show significantly increased criminal activity during the years of Marcellus Shale development, with no change in serious crimes (a rate of 5.8 in 2001–2007 and 5.7 in 2008–2010) and only slight increases for minor crimes (19.0 and 19.5, respectively), DUI (3.4 and 3.7, respectively) and drug abuse violations (1.6 and 2.0, respectively) (Table 7). In Lycoming County, only the arrest rate for DUI was higher during active well development (4.5 and 5.1 arrests per 1,000 residents, respectively). The arrest rates for serious crimes remained the same between the two time periods (8.9 per 1,000 residents in both time periods); the arrest rates for minor crimes and drug abuse violations were lower during active well development in Lycoming County. The arrest rates for all crime types in both Greene and Washington counties were lower during the period of active well development.

New Cases Filed in Criminal, Civil, and Magisterial Courts

The Administrative Office of the Pennsylvania Courts records caseloads within the criminal, civil, family, orphan, and magisterial district court systems through the Pennsylvania Unified Judicial System (UJS). The UJS records new cases filed in a given year by county, indicating the level of criminal activity from the preceding time period (stretching from a few days to a few years) and the load carried by agencies and offices that investigate, prosecute, defend, and adjudicate offenders. It does not provide a measure of the total case load in the system because it does not include cases carried forward from previous years.

As indicated in Table 7, the rates of criminal cases were slightly higher during the period of active well development (2008–2010) than in previous years in three of the four study counties (Bradford, Greene, and Washington). It should be noted though that the rates for each region and for the state were also higher during those years. The rates for civil cases were substantially higher in all study counties; however, the rates also were increasing in other Pennsylvania counties, as indicated by the regional analyses and statewide figures. All four study counties experienced increased rates of traffic violations during the years of active well development, increases for three of the four counties that were larger than those experienced by counties in the surrounding regions and the state.

These findings suggest that some crimes may be affected by Marcellus Shale development in certain locations. However, the exact mechanisms by which this occurs are unknown. Although there is some evidence of increased risky behaviors (e.g., alcohol, drug abuse, etc.) among those employed in the oil and mining industries (Parkins and Angell 2011; Lockie et al. 2009), establishing a link to the natural gas industry is complicated. Further, increased crime rates may be related to increased reporting when communities change (Freudenburg and Jones 1991; Krannich et al. 1985). Changes in enforcement also may lead to shifting reporting patterns (Ruddell 2011). Finally, other developments that occurred during the decade of this study (e.g., the recession) also may have influenced trends in crime.

Perceptions of Community Change

Focus group participants raised concerns about how Marcellus Shale activity is affecting their quality of life and the interactions among community members. They related these concerns to their perceptions of the industry itself and community conflict about the development. Some also discussed how their concerns affect their perceptions of their communities.

"There's No Middle Ground": Community Divisions over Natural Gas

Focus group participants described how they thought Marcellus Shale development has created divisions between some residents. Participants perceived divisions based on disagreements about the prioritization of economic benefits over impacts on the environment. A participant (local government official?) explained the divide this way:

> There's a split between the people. There's, "I don't wanna upset the environment," and there's people … in this town who are greedy and wanna get the money in from the gas. Then there's people that look at it the way everyone else here is talking. "Get out. You're ruining our place. There's no more trees. There's no more animals. It's urbanizing, and that's not what everyone wants here." Then again, there is people with a lot of land that want the money and want what comes with it.

Questions about the distribution of benefits and costs of Marcellus Shale development have created another division. Some economic benefits accrue to local residents through increased business activity, new jobs, and leasing and royalty income. For example, one participant explained that funds generated from gas leasing were used by farmers to reinvest in their operations. This has benefited farmers for whom purchasing new equipment has long been outside of their budget. An owner of an agribusiness explained:

> Oh, it's been a boon to us…. Best thing that ever happened….People's got money and they can spend it. They're either gonna buy something that they can touch or they got to pay the government. It's that simple….There's a lot of rusty machinery here and as soon as you get that kind of a windfall, you would like to have something that's comfortable and shiny.

This increased income can alleviate economic hardship for many community members. This improvement is reflected more than just physical changes:

> [The shale gas industry has] put a lot of money in this county… if you drive [to] the eastern end of the county … it was like Appalachia. I mean the people who lived there, their farms were poor, … there's four or five cars and they make one run out of four or five, and their lawnmower's laid there and the brush was growing up and the house was half built and three kinds of siding on it…. You drive down through there now there's nice manicured yards, they got one decent new car, the junk is gone, the house is fixed up, roof and all one color. I'm not being sarcastic or smart but it made those people proud to have a few dollars that they could better themselves.

Other focus group participants described many problems associated with development, problems not offset by economic gain:

> The complaint's on the truck traffic that's too fast, it's got the road blocked. …When they cut that turn, they run through my yard. There are a lot of people who are complaining. I tend to find that the people who probably do the most complaining are probably the people who did not benefit financially from a lease.

The implications of development for individual and community health and changes to the local population, environment, and amenities are a source of conflict.

These implications, however, are uncertain:

> It seems, as with anything controversial… You can have [name of person] and her farm who says, "My cows are glowing, and the water is on fire." You live next door, [name of participant],

> and you're saying, "I swim in the water. I drink the water. My baby takes a bath in the water." There's no – everything is so up in the air about the health issues.

Focus group participants also reported worries about what new residents might bring to the local community:

> I'm very concerned about the social fabric of the town breaking down. On any given street now, homes are lived in now by gas people.... There's no commitment to the community as far as they're concerned. Those that are renting, they're not really contributing to the tax base either. If you walk around [town], [it's] starting to look like a third-world community. There's no care. People don't keep up the properties. If anything, there's property destruction.

The trade-offs of natural gas development were brought up frequently by focus group participants. They frequently used the phrase *it's a double-edged sword.* Other comments included, *I think you're going to find as many negatives as positives.* Others responded that *it's too soon to tell* or *"ask us in ten years* when asked about whether gas drilling will be good or bad for their regions.

"Not the Same Feeling": Attachment to Place

Focus group participants described how the social and environmental changes they perceived to be occurring affected their feelings about the places in which they lived.

> I love being in the woods, and it's almost like ever since the industry came into the county, there's just less woods around. The land is just pretty much gettin' tore up, so to speak; they put it back, but it's just there's almost not the same feeling in the county as what there was five, six years back.

They used many phrases to describe the impacts on the landscape and how those changes affected their emotional and physical experiences in their favorite places. One participant described the changes to the landscape as *"scars."* Another said:

> One of my favorite things in the whole world when you're having a rough week is to go to [name of] Lake in my canoe and sit out there and fish, and I have a big, huge, loud, noisy well out there. ... sometimes it breaks my heart that the scenery has been just transformed.

For some focus group participants, these threats to their homes inspired political action, advocacy that sometimes comes at a cost:

> If I want to sustain life here, and I want my kids to grow up here in that same type of atmosphere that I did, that's not going to happen just sitting there ... you really have to go to out and strive to achieve that... a lot of people who have joined in with us for that certain goal ... have dropped to the wayside because it's so stressful.

Focus group participants reflected on the balance of short-term economic benefits and long-term risks to their communities and natural resources. *What about the people who actually like living here and want to live here? I feel like I have a split personality, or maybe I want the benefits but I don't want the challenge. I want my cake and eat it, too.* This uncertainty has led to conflicting perspectives about the future. For some focus group participants, the economic opportunities might bring young people back. Others told stories of young people who now want to leave the area or who, if they had left, would never want to return now.

> The …changes in the last 4 or 5 years is drastic … I've always considered myself born and raised in the western section of the county…. You couldn't convince me to consider being anywhere else, but I've told people for the last few years, if I went home tonight and my wife said let's move, and as long as it's a place that I would accept, I'd start looking tomorrow…. the quality of life in [the] county has changed. I'm not saying it's destroyed, I'm saying it's changed – it's not what it used to be…. it'll never be the same again and that's a fact of life.

Others also expressed a desire to leave the community because of the changes: *I don't own the rights on my farm…but I'll tell you this, if I did, I would have sold them. Me and my two kids would probably have moved someplace else.*

Conclusion

The descriptive analyses of quantitative data presented here suggest that the impacts of Marcellus Shale activity on the four study counties are mixed, although they are often hard to distinguish from broader, long-term trends. The findings indicate that some counties have experienced change across some of these sectors but that those changes are conditioned by the level of rurality of the county prior to Marcellus Shale development, geographic region, and preexisting trends in population and economic change. While the secondary data reveal relatively small aggregate changes, the qualitative data suggest substantial community change as well as trepidation about the future of the communities. Together, the research reveals the specificity of experiences within places, the importance of place to residents, and how perceptions of community well-being can be affected by rapid change.

Acknowledgments This research was sponsored by a grant from the Center for Rural Pennsylvania, a legislative agency of the Pennsylvania General Assembly. The Center for Rural Pennsylvania is a bipartisan, bicameral legislative agency that serves as a resource for rural policy within the Pennsylvania General Assembly. Information contained in this report does not necessarily reflect the views of individual board members of the Center for Rural Pennsylvania. The project team gratefully acknowledges several individuals who contributed to the project through background research, data collection, and analysis, including Matthew Filteau, Mark Leach, Kylie Davis, Kirsten Hardy, and Kaitlyn Chajkowski. We also thank the members of our advisory committees who provided us with important insights at key points in the project and gave us access to data sets as well as the community members and county commissioners who shared their experiences through focus groups and interviews. We greatly appreciate their time, assistance, and insights.

References

Adams R, Kelsey TW (2012) Pennsylvania dairy farms and Marcellus Shale, 2007–2010. Penn State Cooperative Extension, University Park

Adgate JL, Goldstein BD, McKenzie LM (2014) Potential public health hazards, exposures and health effects from unconventional natural gas development. Environ Sci Technol

Alter T, Ooms T, Brasier K et al (2010) Baseline socioeconomic analysis for the Marcellus Shale development in Pennsylvania. Report to the Appalachian Regional Commission, Washington, DC

American Community Survey: Selected Population Profile (Pennsylvania) (2007) US Census Bureau, Washington, DC. http://factfinder2.census.gov/. Accessed Aug 2012

Boden LI, Biddle EA, Speiler EA (2001) Social and economic impacts of workplace illness and injury: current and future directions for research. Am J Ind Med 40:398–402

Brasier K, Rhubart D (2014) Effects of Marcellus Shale development on the criminal justice system. Report prepared for the Center for Rural Pennsylvania, Harrisburg

Brasier KJ, Filteau MR, McLaughlin DK et al (2011) Residents' perceptions of community and environmental impacts from development of natural gas in the Marcellus Shale: a comparison of Pennsylvania and New York cases. J Rural Soc Sci 26(1):32–61

Brasier K, Davis L, Glenna L et al (2014) The Marcellus Shale impacts study: chronicling social and economic change in the Northern Tier and Southwest Pennsylvania. Report prepared for the Center for Rural Pennsylvania, Harrisburg

Brown RB, Dorius SF, Krannich RS (2005) The boom-bust-recovery cycle: dynamics of change in community satisfaction and social integration in Delta, Utah. Rural Soc 70(1):28–49

Bunker S, Ciccantell P (2005) Globalization and the race for resources. The John Hopkins University Press, Baltimore

Bureau of Economic Analysis (2011) Regional economic accounts: local area personal income. U.S. Department of Commerce. http://www.bea.gov/regional/pdf/overview/regional_lapi.pdf

Bureau of Economic Analysis (2011) Local areas personal income and employment. (2007–2011). http://www.bea.gov/iTable/iTable.cfm?reqid=70&step=1&isuri=1&acrdn=5#reqid=70&step=1&isuri=1

Bureau of Labor Statistics (2014) The recession of 2007–2009. http://www.bls.gov/spotlight/2012/recession/. Accessed 25 Oct 2012

Camasso MJ, Wilkinson KP (1990) Severe child maltreatment in ecological perspective: the case of the Western energy boom. J Soc Serv Res 13(3):1–18

Center for Workplace Information and Analysis (2011) Marcellus Shale fast face. Pennsylvania Department of Labor and Industry, Harrisburg

Centers for Disease Control and Prevention National Center for Health Statistics (2012) NCHS data on injuries. http://www.cdc.gov/nchs/data/factsheets/factsheet_injury.htm. Accessed on 15 July 2012

Centers for Medicare and Medicaid Services (2013a) Community mental health center fact sheet. http://www.cms.gov/Medicare/Provider-Enrollment-and-Certification/CertificationandComplianc/CommunityHealthCenters.html. Accessed 15 July 2012

Centers for Medicare and Medicaid Services (2013b) Federally qualified health center fact sheet. http://www.cms.gov/Outreach-and-Education/Medicare-Learning-Network-MLN/MLNProducts/downloads/fqhcfactsheet.pdf. Accessed 15 July 2012

Centers for Medicare and Medicaid Services (2013c) Rural health clinic fact sheet. http://www.cms.gov/Outreach-and-Education/Medicare-Learning-Network-MLN/MLNProducts/downloads/RuralHlthClinfctsht.pdf. Accessed 15 July 2012

Corbin JM, Strauss AL (2008) Basics of qualitative research. Sage, Thousand Oaks

Creswell JW (2013) Research design: qualitative, quantitative and mixed methods approaches. Sage, Thousand Oaks

Davis LA, McLaughlin D, Uberoi N (2014) The impact of Marcellus Shale development on health and health care. Report prepared for the Center for Rural Pennsylvania, Harrisburg

Davis KR, Kelsey TW, Glenna LL et al (2014) Local governments and Marcellus Shale development. Report prepared for the Center for Rural Pennsylvania, Harrisburg

Emergency Medical Services (2013) National Highway Transportation Safety Administration, Washington, DC. http://www.ems.gov/mission.htm. Accessed 2 July 2013

England JL, Albrecht SL (1984) Boomtowns and social disruptions. Rural Soc 46:230–46

Farren M, Weinstein A, Partridge M, et al (2013) Too many heads and not enough beds: will shale development cause a housing shortage? The Swank Program in Rural–urban Policy, The Ohio State University, Columbus. http://go.osu.edu/shale_housing_rpt

Ferrar KJ, Kriesky J, Christen CL et al (2013) Assessment and longitudinal analysis of health impacts and stressors perceived to result from unconventional shale gas development in the Marcellus Shale region. Int J Occup Environ Health 19(2):104–112

Finkel ML, Selegean J, Hays J et al (2013) Marcellus Shale drilling's impact on the dairy industry in Pennsylvania: a descriptive report. New Solut 23(1):189–201

Food and Water Watch (2013) The social costs of fracking: a Pennsylvania case study. http://documents.foodandwaterwatch.org/doc/Social_Costs_of_Fracking.pdf. Accessed 22 May 2014

Freudenberg WR (1984) Boomtown's youth: the differential impacts of rapid community growth on adolescents and adults. Am Sociol Rev 49(5):697–705

Freudenburg WR (1984) Differential impacts of rapid community growth. Am Sociol Rev 49:697–715

Freudenburg WR (1992) Addictive Economies: Extractive Industries and Vulnerable Localities in a Changing World Economy. Rural Sociology 57(3):305–332

Freudenburg WR, Frickel S (1994) Digging deeper: mining-dependent regions in historical perspective. Rural Soc 59:266–288

Freudenburg WR, Jones RE (1991) Criminal behavior and rapid community growth: examining the evidence. Rural Soc 56(4):619–645

Freudenburg WR, Wilson LJ (2002) Mining the data: analyzing the economic implications of mining for nonmetropolitan regions. Soc Inq 72(4):546–575

Freudenburg WR, Bacigalupi LM, Landoll-Young C (1982) Mental health consequences of rapid community growth: a report from the longitudinal study of boomtown mental health impacts. J Health Hum Resour Adm 4(3):334–351

Frey J (2012) The future of the Pennsylvania dairy industry with the impact of natural gas. Center for Dairy Excellence, Harrisburg

Gilmore JS (1976) Boom towns may hinder energy resource development: Isolated rural communities cannot handle sudden industrialization and growth without help. Science 191:535–540

Glenna L, Babbie K, Kelsey TW et al (forthcoming) Establishing a baseline for measuring agricultural changes related to Marcellus Shale development. Report prepared for the Center for Rural Pennsylvania, Harrisburg

Gramling R, Freudenburg WR (1990) A closer look at 'local control': communities, commodities, and the collapse of the coast. Rural Soc 55(4):541–558

Harper JA (2008) The Marcellus Shale-an old "new" gas reservoir in Pennsylvania. Pennsylvania Geol 38:2–13

Headwaters Economics (2009) Fossil fuel extraction as a county economic development strategy: are energy-focusing counties benefiting? Headwaters Economics, Bozeman

Jacobson M, Kelsey TW (2011) Impacts of Marcellus Shale development on municipal governments in Susquehanna and Washington counties, 2010. Penn State University Cooperative Extension Marcellus Education Fact Sheet, University Park

Jacquet JB, Kay DL (2014) The unconventional boomtown: updating the impact model to fit new spatial and temporal scales. J Rural Commun Devel 9(1):1–23

James A, Aadland D (2011) The Curse of the natural resources: an empirical investigation of U.S. counties. Resour Energy Econ 33:440–453

Kelsey TW, Adams R, Milchak S (2012a) Real property tax base, market values, and Marcellus Shale: 2007–2009. CECD Research Paper Series. http://cecd.aers.psu.edu

Kelsey T, Hartman W, Schafft KA, et al (2012b) Marcellus Shale gas development and Pennsylvania school districts: what are the implications for school expenditures and tax revenues? Penn State Cooperative Extension Marcellus Education Fact Sheet, University Park

Kelsey TW, Hardy K, Glenna L et al (forthcoming) Local economic impacts related to Marcellus Shale development. Report prepared for the Center for Rural Pennsylvania, Harrisburg

Kinchy A, Perry S, Rhubart D et al (2014) New natural gas development and rural communities in the United States. In: Bailey C, Jenson L, Ransom E (eds) Rural America in a globalizing world: problems and prospects for the 2010s. West Virginia University Press, Morgantown

Kohrs E, Dean V (1974) Social consequences of boom town growth in Wyoming. Paper presented at the Rocky Mountain American Association of the Advancement of Science Meeting, Laramie, Wyoming. 24–26 Apr1974

Kowalski L, Zajec G (2012) A preliminary examination of Marcellus Shale drilling activity and crime trends in Pennsylvania. Justice Center for Research, Pennsylvania State University, University Park. http://justicecenter.psu.edu/research/documents/MarcellusFinalReport.pdf

Krannich RS (2012) Social change in natural resource-based rural communities: the evolution of sociological research and knowledge as influenced by William R. Freudenburg. J Environ Stud Sci 2(1):18–27

Krannich RS, Greider T, Little RL (1985) Rapid growth and fear of crime: a four-community comparison. Rural Soc 50(2):193–29

Lockie SM, Ranettovich V, Petkova-Timmer J et al (2009) Coal mining and the resource community cycle: a longitudinal assessment of the social impacts of the Coppabella coal mine. Environ Impact Asses 29(5):330–339

Lycoming Department of Planning and Community Development (2012) The impacts of the Marcellus Shale industry on housing in Lycoming County. http://www.lyco.org/Departments/PlanningandCommunityDevelopment.aspx

McDermott-Levy R, Kaktins N (2012) Preserving health in the Marcellus region. Pennsylvania Nurse September 67(3):4–10

McLaughlin D, DeLessio-Parson A, Rhubart D (2014a) Housing and Marcellus Shale development. Report prepared for the Center for Rural Pennsylvania, Harrisburg

McLaughlin D, Rhubart D, DeLessio-Parson A et al (2014b) Population change and Marcellus Shale development. Report prepared for the Center for Rural Pennsylvania, Harrisburg

Ooms T, Tracewski S, Wassel K et al (2011) Impact on housing in Appalachian Pennsylvania as a result of Marcellus Shale. The Institute for Public Policy and Economic Development, Wilkes-Barre. http://www.institutepa.org/PDF/Marcellus/housing11.pdf

Parkins JR, Angell AC (2011) Linking social structure, fragmentation, and substance abuse in a resource-based community. Commun Work Family 14(1):39–55

Pennsylvania Department of Environmental Protection Office of Oil and Gas Management: Wells Drilled by County (2014). Pennsylvania Department of Environmental Protection, Harrisburg. http://www.portal.state.pa.us/portal/server.pt/community/oil_and_gas_reports/20297. Accessed 21 May 2014

Pennsylvania Department of Public Welfare (2013) Achieves of food stamps and cash stats. http://listserv.dpw.state.pa.us/ma-food-stamps-and-cash-stats.html. Accessed 12 July 2013

Pennsylvania Department of Revenue (2013) Tax Compendium, 2007–12. http://www.portal.state.pa.us/portal/server.pt/community/reports _and_statistics/17303/tax_compendium/602434

Perry S (2012) Development, land use, and collective trauma: the Marcellus Shale gas boom in rural Pennsylvania. Culture Agric Food Environ 34(1):81–92

Ruddell R (2011) Boomtown policing: responding to the dark side of resource development. Policing 5(4):328–342

Rural Sociological Society (1993) Persistent poverty in rural America. Westview Press, Boulder

Schafft KA, Biddle C (2014) Youth perspectives on Marcellus Shale gas development: community change and future prospects. Report prepared for the center for rural Pennsylvania, Harrisburg

Schafft KA, Borlu Y, Glenna L (2013) The relationship between Marcellus Shale gas development in Pennsylvania and local perception of risk and opportunity. Rural Soc 78(2):143–166

Schafft KA, Glenna LL, Green B et al (2014a) Local impacts of unconventional gas development within Pennsylvania's Marcellus Shale Region: gauging boomtown development through the perspectives of educational administrators. Soc Natur Resour 27(4):389–404

Schafft KA, Kotok S, Biddle C (2014b) Marcellus Shale gas development and impacts on Pennsylvania schools and education. Report prepared for the Center for Rural Pennsylvania, Harrisburg

Seyfritt CL, Sadler-Hammer NC (1988) Social impact of rapid energy development on rural youth: a statewide comparison. Soc Natur Resour 1(1):57–67

Social Explorer Tables: Census 2000, Census 2010, American Community Survey (2005–2007 and 2009–2011) three-year estimates (2011) Social explorer. http://www.socialexplorer.com/. Accessed 12 July 2013

Steinzor N, Subra W, Sumi L (2013) Investigating links between shale gas development and health impacts through a community survey project in Pennsylvania. New Solut 23(1):55–83

Stokowski PA (1996) Crime patterns and gaming development in rural Colorado. J Travel Res 34:63–69

The Unified Judicial System of Pennsylvania (2014). Research and Statistics Records: Common Please Case Management System & Magisterial District Judge Data. http://www.pacourts.us/news-and-statistics/research-and-statistics/

US Census Bureau (2010) Table 1. Intercensal estimates of the resident population for counties of Pennsylvania, April 1, 2000 to July 1, 2010. https://www.census.gov/popest/data/intercensal/county/CO-EST00INT-01.html

U.S. Census Bureau (2013) Small area health insurance estimate (SAHIE). http://www.census.gov/hhes/www/sahie/. Accessed 12 July 2012

U.S. Census Bureau, Population Estimates (2013) http://www.census.gov/popest/. Provided on 15 July 2013

U.S. Census Bureau Quick Facts (2013) http://quickfacts.census.gov/qfd/states/42/42015.html

U.S. Department of Health and Human Services (2012) Access to care. http://www.healthypeople.gov/2020/topicsobjectives2020/overview.aspx?topicid=1. Accessed 12 July 2012

U.S. Department of Housing and Urban Development (HUD) (2014) Affordable housing. http://portal.hud.gov/hudportal/HUD?src=/program_offices/comm_planning/affordablehousing/

United States Department of Agriculture's National Agricultural Statistics Service (2013) Farms, Land in Farms, and Livestock Operations: 2012 Summary. http://usda01.library.cornell.edu/usda/current/FarmLandIn/FarmLandIn-02-19-2013.pdf

United States Department of Justice. Federal Bureau of Investigation. Uniform Crime Reporting Program Data [United States]: County-Level Detailed Arrest and Offense Data, 2008. ICPSR27644-v1. Ann Arbor, MI: Inter-university Consortium for Political and Social Research [distributor], 2011-04-21. doi:10.3886/ICPSR27644.v1

Wilkinson KP (1985) Community development in rural America: sociological issues in national policy. J S Rural Sociol (3)1

Williamson, J, Kolb B (2011) Marcellus natural gas development's effect on housing in Pennsylvania. Lycoming college center for the study of community and the economy, Williamsport, Pennsylvania. http://www.housingalliancepa.org/sites/default/files/resources/Lycoming-PHFA%20Marcellus_report.pdf

Origins and Consequences of State-Level Variation in Shale Regulation: The Cases of Pennsylvania and New York

Ilia Murtazashvili

Abstract The combination of hydraulic fracturing and horizontal drilling unlocked the economic potential of shale gas in the United States. However, the regulatory response to shale gas has varied substantially. This chapter considers the political economy of Marcellus Shale, focusing on differences in regulatory responses in Pennsylvania and New York. It suggests that the regulatory response in Pennsylvania is "efficient" and that different responses can be explained by features of politics, rather than geography, relative prices, or institutions. The chapter concludes by considering the benefits and costs of federal regulation of fracking. Although states have varied substantially in their response to fracking, decentralized governance of shale gas has many benefits and few discernible costs.

Introduction

It has long been known that the United States is rich in shale gas. Until recently, the shale was not economically valuable since the gas cannot be profitably extracted using conventional downward drilling techniques. The combination of hydraulic fracturing (or "fracking") and horizontal drilling unlocked the vast potential of shale gas for these states. Neither fracking nor horizontal drilling is a new technology, although the consequences of combining these technologies are similar to a new technology of extraction of shale gas (Fitzgerald 2013).

One of the remarkable features of shale extraction is the rapid response of economic actors to new economic opportunities. According to "efficiency" perspectives on institutional change, economic actors respond to opportunities for wealth creation as long as the state provides a basic system of private property rights (Demsetz 1967; Barzel 1989). Although political and regulatory conflict often

I. Murtazashvili (✉)
Graduate School of Public and International Affairs, University of Pittsburgh,
3810 Wesley W. Posvar Hall, Pittsburgh, PA 15260, USA
e-mail: imurtaz@pitt.edu

W.E. Hefley, Y. Wang (eds.), *Economics of Unconventional Shale Gas Development*, Natural Resource Management and Policy 45,
DOI 10.1007/978-3-319-11499-6_8

undermines opportunities for economic actors to take advantage of new economic opportunities (Libecap 1989), the boom in the production of shale gas suggests few obstacles to taking advantage of these economic opportunities.

North (1990) suggests that the institutional matrix of society, in particular the structure of property rights, is the key to understanding the extent an economy takes advantage of new economic opportunities. The property system of the United States provided a nearly ideal foundation for the shale boom. Part of the reasons is because the property regime governing minerals has been evolving for over two centuries. The Land Ordinance of 1785 and the Northwest Ordinance of 1787 established a framework to survey land for the purposes of establishing private property rights to land (North and Rutten 1987; Mittal et al. 2011). Although the land ordinances defined a private property system for surface land, these ordinances also established a substantial realm of state ownership of mineral lands. State ownership of mineral land eventually would come into conflict with the demands of individuals seeking to extract the mineral resources during the "gold rush" in the American west. From 1848, the discovery of gold in California, until 1866, the federal government technically owned mineral land and the mineral rights (Umbeck 1981). Despite formal government ownership, individuals contracted for property rights on their own, devising governance systems known as mining districts to enforce private property rights to mineral land (Murtazashvili 2013). Eventually, the federal government strengthened the property system, first with the California Land Claims Act of 1851, which helped clarify features of landownership, and in 1866, when the Mining Act decentralized mineral rights and formalized local mining codes (Libecap 1989; Clay 1999; Clay and Wright 2005).

Although there was substantial conflict between the demand for private ownership of minerals and the desire of the state to profit from its vast mineral lands, by 1866, the system was well on its way to decentralization of ownership of mineral lands. By that time, there were three property regimes governing minerals. One is fee simple ownership, whereby the surface owner retains rights to both surface land and the mineral estate (Ellickson 1993). There are also split estates, whereby the surface and mineral rights are severed, which is common where individuals wanted to sell surface land and retain mineral rights. In such cases, mineral right owners usually have rights of "reasonable" access to mineral land. There is also, in some situations, state ownership of the mineral estate, for example, on some state-owned lands (although on some state-owned lands, the state only owns surface land, and individuals own mineral rights).

The key with respect to shale is that property rights to mineral were clearly allocated and secure. Regardless of whether the property system is state ownership, fee simple, or split estate, it is fairly easy to discern who has the rights to shale gas (although figuring out who actually owns mineral rights involves some legwork, which is why landmen feature so prominently in the shale boom). As a consequence, gas companies were able to lease land fairly quickly. Much like Coase (1960) predicted, when the transaction costs are low and the state provides a system of private property rights, property rights were allocated efficiently to their highest-value use. Although *Promised Land*, the Hollywood account of the shale boom, is critical of the process of allocating land, it is quite remarkable that the contracting parties were able to reallocate property rights so quickly in response to new economic opportunities.

As a result of this favorable institutional environment, shale production has increased from about 1 % to a third of output. In addition, there has been improvement in jobs (Weber 2012). Despite efficiency in the leasing process, and the dramatic increase in shale production, there has been tremendous variation in state-level regulation in response to opportunities and challenges presented by refinements in technology to extract shale gas. One of the sharpest differences is in the neighboring states of Pennsylvania and New York. Although these states share a border and expect similar benefits from shale production, the response to shale gas differed dramatically: Pennsylvania responded to new opportunities by encouraging shale production, while New York has had a de facto moratorium on shale production since 2009. This chapter considers the different regulatory responses in these states, the reasons for these differences, and the extent to which federalism, whereby states have substantial autonomy to regulate fracking, is appropriate in the case of shale regulation.

There are several reasons why variation in these states is interesting from a political economy perspective. One is because they represent two very different responses to new economic opportunities. Pennsylvania is representative of a legal regime that promotes fracking, while the response in New York, which is the contrast case, is representative of a much more cautious response to fracking.[1] As such, these cases are useful reminders of the importance of politics in understanding how economies respond to new economic opportunities.

Pennsylvania and New York are also a useful comparison from a research design perspective. Once the "efficiency" of different responses is identified, it is important to understand why responses are efficient in some cases but inefficient in others (Riker and Weimer 1995; Weimer 1997). Explaining the extent to which rules reflect efficiency considerations requires a theory of institutional change, although establishing which mechanisms drive the process of institutional change is often difficult. Some of the major explanations for variation in rules and regulations include geography, changes in relative prices, institutions, and politics. Intuitively, "politics" helps explain variation in shale policies, although it is important to at least make a case that politics is a causal mechanism explaining variation in shale policies. As Pennsylvania and New York share a common border, similar geography, similar expected gains from shale production, and similar property institutions, it is less likely that geography, changes in relative prices, and property institutions explain variation in shale policies. This feature of the research design provides a useful opportunity to understand whether politics and group conflict explain differences in these policies since we can rule out some of the major competing explanations for variation in the response to these policies.

[1] Much of the Marcellus Shale lies atop the deeper Utica Shale. As the Marcellus is more profitable to extract, it has been more extensively mined at this point.

This chapter offers several conclusions regarding state-level regulation of Marcellus Shale. First, despite popular accounts that Pennsylvania encouraged a Wild West situation with respect to fracking, the regulatory response has been quite rapid and comprehensive on several generally accepted dimensions of efficiency. Second, variation in political actors appears to be a compelling explanation for differences in state-level Marcellus policies. Third, there is a compelling case to be made that federalism is an appropriate institutional solution to the issue of fracking. Unlike several recent studies that use variation in regulation as evidence that there is a need for a larger role for the federal government (e.g., Wiseman 2009; Richardson et al. 2013; Warner and Shapiro 2013), this chapter clarifies that the mere fact of substantial regulatory variation does not imply a "fractured" or inadequate regulatory regime. Rather, a strong case for regulatory federalism can be made due to heterogeneity of preferences for fracking, the importance of policy experiments with a new technology, and the regulatory capacity of the local units (both the states and local governments), as well as because of few discernible interstate externalities associated with fracking.

The chapter is organized as follows. The next section reviews the regulatory responses in Pennsylvania and New York. Section "Is Act 13 "Efficient?" assesses the regulatory responses on several dimensions to understand the extent to which state regulations are efficient. Section "Explaining Regulatory Response" seeks to explain the regulatory response, in particular the extent to which geography, changes in relative prices, institutions, and politics are plausible explanations for variation in shale policies. Section "Federalism and Fracking" reviews the case for federalism of fracking regulation. The conclusion suggests several areas for future research.

Regulatory Response

The rapid reallocation of property rights and the dramatic increase in shale production suggest that the economy has responded rapidly to new economic opportunities. In both Pennsylvania and New York, there was a fairly orderly process of leasing land, one that set the stage for shale production. However, Pennsylvania enacted legislation that increased opportunities for shale production, while New York responded with an extended study of the consequences of fracking that serves as a de facto moratorium despite substantial leasing activity. This section considers the key features of the legislation response in these states.

Pennsylvania

The major legislation regarding shale in Pennsylvania is Act 13, which was signed by Governor Tom Corbett in 2012. By the time of Act 13, there had already been substantial activity regarding shale. Exploration began in earnest in 2005–2006 in

Pennsylvania. Shortly thereafter, companies began to secure lease rights to land. The first companies were often smaller companies whose purpose was to pool leases rather than to engage in fracking. These companies began pooling leases, speculating that the pooled leases would be valuable later on as larger companies would come in to actually drill wells. The companies do not bargain for leases themselves; rather, they rely on landmen to negotiate contracts, as well as figure out who owns mineral rights to land. Discerning ownership can be a challenging task since the property rights to surface and subsurface land are often divided in the United States. Therefore, the challenge of leasing land often requires figuring out who actually owns the mineral rights. Landmen, in this regard, are entrepreneurs who help to increase social surplus by bringing contracting parties together.

One of the central mechanisms of institutional change in economics is change in relative prices. As relative prices change, individuals have incentives to establish new institutions (North and Thomas 1973; North 1981). Once institutions are established, changes in relative prices motivate reallocation of property rights. The rapid emergence of the shale sector reflected not only the strength of property rights but the rich traditional of conventional oil and gas law in Pennsylvania that could be adapted for the purposes of unconventional oil and gas exploration and production.

Although gas companies and their landmen brought the promise of shale wealth to rural Pennsylvania, there was almost immediate opposition to fracking from local governments. Several municipalities in Pennsylvania responded to the nascent shale boom with regulations prohibiting fracking. The first municipality in the country to ban fracking was Pittsburgh, Pennsylvania. This was perhaps ironic given Pittsburgh's role in the coal and steel industry. The city council unanimously voted to ban fracking in the city based on concerns about its environmental consequences, although at the time only about 1 % of land was under lease with gas companies. The ban, which was drafted by the Community Environmental Legal Defense Fund and spearheaded by council member Doug Shields, was based both on concern about fracking and on the political theory that local governments have a right to self-governance (Smydo 2010). Subsequently, the Pittsburgh City Council voted to put the regulation to a referendum in order to change the home rule charter of the city and make it more difficult for future city governments to rescind it (Smydo 2011). Ultimately, then-mayor Luke Ravenstahl decided not to sign the referendum bill, and so it was not put to a referendum (McNulty 2011). The commissioners of Allegheny County, which includes Pittsburgh, also considered a ban, and at the time 7 % of land was under lease, but ultimately did not act on it (Smydo and Barcousky 2011). Several other local governments banned fracking, including Wilkinsburg, which borders Pittsburgh, as did several other communities.

There were several reasons why local governments adopted bans on fracking. One issue was concern about the economic impact of shale. Many communities had seen extractive industries come and go and did not believe that fracking was a sustainable future. The communities seeking to ban fracking also expressed a belief in local self-determination, which is the right of local governments to determine their collective futures. There was also substantial uncertainty about the environmental and ecological consequences of fracking.

Yet at the core, the bans on fracking reflected a belief that legislation at the local level was necessary because state regulations were poorly equipped to handle the challenge of fracking. There was no shortage of popular and academic support for the argument that the state had few regulations in place. Shale production was described as a Wild West of laissez-faire regulations more common of the nineteenth century (Rabe and Borick 2012; Revkin 2013). Others have described shale as a "blind rush" due to the perception it was occurring without much in the way of oversight (Schmidt 2011). Another study describes shale production as "untested waters," suggesting caution with shale production despite acknowledging that the EPA had already conducted several large-scale studies of the consequences of fracking (Wiseman 2009).

Although there are obvious differences between conventional and unconventional gas production, such as much more of a concern with groundwater contamination with fracking (Holahan and Arnold 2013), states such as Pennsylvania have a fairly powerful regulatory regime in place due to a long legacy of extraction of conventional oil and gas resources. There is also a large body of law that has dealt with issues of leasing, pooling, and contracting for shale rights, as well as the authority of the state to preempt local regulation (Pifer 2010a, b). This suggests that the legal environment was less a Wild West than one for which regulations exist, but had to be modified in response to differences in conventional and unconventional extraction.

As the state had taken the lead in regulating conventional oil and gas, the logical place to reform regulations governing shale is at the state level. Governor Tom Corbett worked modify existing state-level rules, rather than pursuing a new policy of decentralizing regulatory authority to local governments. Governor Corbett appointed an advisory committee to study Marcellus Shale in 2011, which after a year of public meetings and study of the issues produced a report that ultimately became Act 13, which amended the Oil and Gas Act of 1984 (Governor's Marcellus Shale Advisory Commission 2011). Act 13 has several major provisions, including preempting local zoning and imposed an impact fee. The impact fee was controversial, as some lawmakers supported (and still support) a severance tax. In addition, there was concern that the preemption of local authority violated the *Home Rule Charter* that provides local communities with zoning authority.

Act 13 also established a model zoning ordinance, one that specified setback requirements for local governments. This was important since around 40 % of communities where fracking occurs do not actually have a zoning requirement (Colaneri 2014). In addition, the impact fee included a set of authorized uses of fees that was designed to ensure funds were used to make sure communities used resources to fund Marcellus-related improvements to local infrastructure.

Several communities challenged Act 13 as a violation of the principle of the *Home Rule Charter* that provides for local self-governance. In December 2013, the Pennsylvania Supreme Court ruled 4–2 to overturn the zoning restrictions. They also overturned the setback provisions, which meant that currently there is no clear rule governing setbacks in areas that have no zoning requirements (Cusick 2014). The Corbett administration has appealed, as has the state Department of Environmental Protection, the latter objecting to the court overturning the setback provisions (Hopey 2014).

New York

Vast quantities of shale gas also lie below the surface in New York. Just as in Pennsylvania, many landowners in New York signed leases with gas companies in anticipation of a shale boom that was already occurring in other parts of the country. In this regard, New York is also an example of thorough and rapid reallocation of property rights in response to new economic opportunities as private parties leased land to its new highest-value use.

The similarity ends at the leasing process, however. The state of New York from the outset pursued quite a different policy toward shale. In response to health concerns, Governor Andrew Cuomo authorized in 2009 a health impact study to understand the environmental consequences of fracking, one that has had effects similar to a moratorium on fracking and has been met with substantial opposition, in part due to concerns that the impact study has already been completed, yet its findings delayed (Hakim 2013; McKinley 2013a). Despite opposition, the de facto moratorium remains in place as of 2014.

Although the state imposed a de facto ban on fracking, there has been substantial political conflict over fracking at the local level in New York. Similar to Pennsylvania, the state of New York has a home rule that provides substantial authority to local governments to regulate economic activities. The first community in New York to challenge fracking directly was Dryden, a rural community of about 15,000 in Tompkins County that boasts farms and horse ranches and several small businesses. In August 2011, the town's Board of Supervisors voted to ban hydraulic fracturing after a lengthy lobbying debate. It was supported by the Dryden Resources Awareness Coalition. Some were already offered several thousand dollars an acre to lease the land, which gives an idea of the stakes of the ban (McKinley 2013b). In all, around 10 % of local governments in areas with the potential for shale production have banned fracking or have attempted to do so (Arnold and Holahan 2014).

As one expects, these bans on fracking met substantial opposition. The de facto ban on fracking harms landowners who signed leases and whose royalty payments depend on shale production. The de facto ban also adversely affects those who wish to sign leases since fracking influences the value of their lease rights. In addition, industry had already invested in securing lease rights. Perhaps unsurprisingly, the regulations at the local level were challenged by the Independent Oil and Gas Association of New York. The regulation in Dryden was also challenged by Norse Energy Corporation USA in court.

The issue in the legal cases is whether state law preempts local authority to ban fracking. Ultimately, the courts ultimately sided with the town of Dryden (Hills 2014). According to these legal decisions, in the absence of a clear intention by the state to preemption local governments, these local governments have the authority to ban fracking. As the political environment at the state level in New York has been less favorable to fracking than in Pennsylvania, there is no clear preemption of local regulations by state oil and gas law, and so it appears that communities are within their rights to ban fracking.

Is Act 13 "Efficient?"

One of the central questions in political economy is the extent to which regulations reflect efficiency considerations. In the narrow sense, an "efficient" policy is conceptualized of as one that promotes wealth creation (Barzel 1989; Knight and North 1997). At the same time, what constitutes an efficient shale policy also depends on the extent to which regulations include provisions to remedy economic externalities. In addition, it is important to understand the extent to which shale regulations provide the state with revenue as well as preserve local autonomy. The discussion that follows compares legislation in Pennsylvania and New York on these different dimensions of efficiency.

Responsive to Wealth

In the Northian tradition of institutional analysis, efficient institutions and policies are those that are responsive to opportunities for wealth creation. In terms of institutional design, Act 13 fares well as far as responsiveness to wealth is concerned. Rather than constrain fracking, Act 13 promotes the development of shale gas. The law effectively limits the ability of local jurisdictions to ban fracking and therefore promoted the ability of gas companies to make production decisions based on economic conditions after they have leased land. There are various ways to measure these gains, including job creation as well as new investment opportunities as individuals receive payments from up-front bonus payments for leasing land and royalties from production. Act 13 thus ensures that production decisions will be made based on changes in relative prices, with gas companies maximizing the values of their pooled leases.

The law in New York, in contrast, is unresponsive to opportunities for wealth creation. The leasing process suggests that there was much to be gained in terms of wealth creation. However, state regulations have prevented shale production and also undermined incentives to sign leases due to uncertainty about the political regime.

Economic Externalities

Economic externalities are market failures that have to be taken into account when considering whether regulations are economically efficient. When there are substantial externalities associated with fracking, there will be fewer gains from contracting for shale property rights.

It is useful to contrast conventional and unconventional oil to understand the economic externalities with shale production. In the broad sense, shale and conventional gas are both common pool resources. The challenge with a common pool resource is that there is tendency for it to be used up in the absence of effective property rights (Libecap 1989; Ostrom 1990; Ostrom 2005). With conventional gas (and oil), a main challenge is racing to extract gas from a common pool, which can be socially wasteful. The solution in this situation is often unitization, or forced pooling, which is where the group decides to impose a collective system of ownership of the well. By establishing collective ownership, each driller is a residual claimant in the production of the well, and so such systems internalize the externalities associated with costly racing to extract oil (Wiggins and Libecap 1985; Libecap and Smith 1999; Libecap and Smith 2002).

The typical problem with unitization is that political conflict may undermine such agreements (Libecap and Wiggins 1985). In contrast, the problem with shale production is that fracking is a nonpoint source of pollution, which refers to situations in which there are multiple individuals contributing to pollution in ways for which it is difficult to assign responsibility for environmental harm (Holahan and Arnold 2013). As such, unitization is not a solution to the environmental problem with shale. Rather, unitization can solve a holdup problem, as forced pooling would require landowners to go along with the drilling plan, but it would not necessarily deal with the issue of groundwater contamination.

The differences between conventional and unconventional oil and gas as far as unitization is concerned do not mean the institutional environment of shale is a regulatory Wild West.[2] As states have long been dealing with issues of nonpoint pollution, the state should have ample ability to deal with the consequences of fracking from a regulatory perspective. In Pennsylvania, the authority to regulate fracking falls to the state Department of Environmental Protection. In addition, it is also important to recognize that the federal Environmental Protection Agency (EPA) has important oversight functions regarding shale.[3]

In terms of institutional design, Act 13 has important provisions for environmental sustainability. One is that it solves an important collective action problem for local governments regarding requirements for how far wells are to be placed from groundwater. As noted above, only about 40 % have zoning for dealing with fracking. Besides many communities lacking zoning, Pennsylvania also has a great deal of local government units, ranking third in the nation in the number of municipalities (2,562 municipalities, including 56 cities, 958 boroughs, 1 town, 93 first-class townships, and 1,454 second-class townships). An important reason for the model zoning ordinance of Act 13 is that there is a potentially huge coordination problem and also because many of these local governments do not have the capacity to regulate shale development.

---

[2] Wiseman (2009), for example, suggests caution with fracking despite EPA oversight in the form of a massive study of fracking and extensive experience with the Barnett Shale, which was extensively mined by even 2009.

[3] See EPA, "A Study of the Potential Impact of Hydraulic Fracturing on Drinking Water Resources—PreliminaryReport,"http://www2.epa.gov/hfstudy/study-potential-impacts-hydraulic-fracturing-drinking-water-resources-progress-report-0

From a design perspective, Act 13 appears to solve a collective action problem. However, shale production has been commencing for long enough that we can determine whether there are widespread economic externalities associated with shale production. Despite a great deal of fear surrounding fracking, there seems to be little evidence of economic externalities associated with shale production. Many of the environmental studies of the consequence of fracking the Marcellus Shale focus on the effects on groundwater. There is much more of an issue with groundwater with the Marcellus than the Barnett in Texas shale as the fracking sites are much closer to homes that rely on groundwater. There was also the highly publicized controversy in Dimock, Pennsylvania, about the fear of methane contamination. The Oscar-nominated documentary "Gasland" suggested that fracking was contaminating groundwater with methane gas, and potentially making it flammable, according to reports from several households in the town (Banerjee 2012). However, the EPA subsequently refuted any link between fracking and groundwater risk in the region (Drajem 2013).

Although there has been much hype surrounding the potential for groundwater contamination, the best scientific studies of fracking appear inconclusive and certainly do not show that there is clear evidence of groundwater contamination. One recent study, with results published in the journal *Science*, shows that while fracking may have some marginal effects on groundwater, establishing a causal relationship is difficult because there are few reliable measures of contamination before fracking commenced (Vidic et al. 2013). There have also been a few studies that address the issue of groundwater contamination indirectly, including by the effect of perceptions of groundwater contamination on property values. A recent study finds fracking reduces property values of homes that depend on groundwater by about 1 % (Muehlenbachs et al. 2014). This finding may be due to actual groundwater contamination, but it may also be a belief that fracking affects groundwater rather than actual groundwater contamination, and so the finding itself is not necessarily evidence of adverse environmental consequences of fracking. In addition, the effects on property values are quite small and may not be permanent. Since the aforementioned study of property values uses data after only about 2 years of shale production, it cannot be used to draw conclusions about the long-run consequences of shale production for property values. There is also some concern about surface water contamination from fracking, although there is also limited evidence of any adverse effect of fracking on surface water (Olmstead et al. 2013).

Besides water contamination, the other key aspect of environmental and ecological sustainability with fracking is release of methane into the atmosphere. The problem with methane escaping during the process is that the fugitive methane can have a large impact on global warming, perhaps more of an impact than coal burning (Howarth et al. 2011). To the extent that fugitive methane is escaping, shale is not exactly environmentally sustainable. There is, however, great debate about just how much of an impact methane has, with some believing that earlier studies overstate the impact (Cathles III et al. 2012; Howarth et al. 2012). In addition, it is critical to recognize that there are two ways to deal with fugitive methane. One is flaring, or lighting up the well to burn the methane. This may seem unfriendly to the environment, yet flaring reduces harmful consequences of fugitive methane by burning it. A second is capping wells, which many companies do. As many gas companies flare or cap (or both), it seems that fugitive methane is not as much of a problem as early studies indicated.

A final issue with sustainability concerns the relationship between shale production and other energy sources, in particular renewable energies. Act 13, in promoting shale, is promoting a fossil fuel. As shale is a fossil fuel, laws that encourage it can be viewed as poorly designed. Indeed, an important argument made against shale gas development is anything that reduces the transition to renewables is "bad" (Grossman 2013). At a minimum, this critique suggests the importance of considering, from a design perspective, the relationship between promotion of shale and the relationship between shale and alternative energy development.

Studies suggesting the institutional environment to deal with fracking is inadequate are too numerous to mention. However, it is not clear that there is much evidence of the inadequacy of regulation in terms of environmental outcomes. If the regulatory regime is inadequate, then we would expect more in the way of unmitigated externalities. In reality, there is little evidence of such environmental problems. Rather, the major legislation promoting fracking in Pennsylvania includes provisions to manage environmental consequences of fracking while also providing the state with revenue to manage the harms associated with fracking (see the discussion on revenue below).

The state of New York does well in terms of environmental preservation as far as groundwater and fugitive methane are concerned. The reason is obvious: there is a de facto ban in place, so there are no externalities associated with shale production. Of course, it is important to consider that the case has been made that natural gas is a better alternative than coal. In this regard, the New York moratorium, because it reduces the use of natural gas, may have environmental costs.

State Revenue

There are several important reasons to consider state revenue when considering natural resource extraction. The resource curse is conceptualized of in terms of outcomes ranging from economic growth to civil war to democracy, as well as the extent to which the state can profit from its natural resource wealth (Ross 1999; Ross 2001; Collier and Hoeffler 2005). Although the resource curse is often associated with the developing world, the resource curse can set in anywhere, including in the US states. The idea is that states with more resources wealth may be worse off because of it (Goldberg et al. 2008). To date, the only study of the shale resource curse has focused on implications for jobs and population (Weber 2014). Yet it is also important to consider state revenue, including because the state's ability to profit from shale influences its ability to provide public goods and deal with the economic externalities associated with shale production.

There are several main options to collect revenue from shale production, in particular impact fees, severance tax, or realizing gains indirectly through increasing property values. Act 13 does not allow local governments to directly tax shale production. Rather, it relies on impact fees and distributes the fees based on a formula.

From a design perspective, Act 13 recognizes that property taxes are an indirect way to realize gains from shale for local government. The impact fees provide a more direct way to ensure local governments benefit than simply relying on increases in property values as a way to increase local government revenue. The impact fees, administered by the state, may also make it less likely that local governments that are not savvy (or are simply shortsighted) impose little or no taxes on shale production. A severance tax, in contrast, could be used for almost anything and would not necessarily be tied to local externalities.

The impact fees have been a boon to many local communities, in particular those with smaller government budgets, as measured by the increase in budget size in response to shale production. The information on the wells and revenues is provided by Public Utilities Commission. Although the number of wells increased by around 25 %, revenue declined in 2012 compared to the year before. In 2011, shale companies contributed $204 million to the state and $198 million in 2012. The amounts are determined by the gas prices, and as production increases, putting downward pressure on prices, revenue may drop. Yearly fees may be as low as $40,000 for a new well, rather than $60,000, and after 15 years, the rates may drop to as low as $5,000 per well (Detrow 2012).

Although it is too soon to understand all the consequences of the impact fees, they appear to provide some local governments with a sizable increase in revenue. A fuller assessment of the consequences of the impact fees would necessarily involve consideration of how the revenues are spent, as well as whether they cover the costs of shale production. As there may be unforeseen costs from fracking, the impact fees may not be set optimally. Nonetheless, it seems clear that in the short run, Act 13 provides communities with a sizable amount of local revenue.

An alternative to the impact fee is a severance tax, which would go to the state, rather than to the communities. One of Governor Corbett's opponents in the 2014 election, Allyson Schwartz, argued that a tax on shale of 5 % would generate billions in revenue. The alternative proposal promised to generate $22 billion in revenue by 2022.[4] Although it is certainly possible to increase taxes on shale production, the impact fee is more directly tied to economic externalities than the severance tax proposal. Nonetheless, either impact fees or a severance tax appears important to ensuring the state acquires revenue, and either an impact fee or a tax, when set optimally, is important to understanding whether the law is efficient. The impact fee appears to fare quite well in terms of efficiency as it is tailored to environmental externalities.

[4]The proposal is found here: http://allysonschwartz.com/wp-content/uploads/Schwartz-Shale-MC13.pdf

New York, at least in the short terms, undermined opportunities for the state to profit from natural resource wealth. Some local governments may be harmed by this. Of course, the shale remains in the ground, so in the long run, there are still opportunities for the state to profit from its shale wealth, and it would be unlikely that the state would allow gas companies to produce without some sort of impact fee or tax (given the less friendly environment to fracking in the state).

Local Autonomy

One of the more controversial aspects of Act 13 is removing authority to ban fracking. Specifically, Act 13 provided a model ordinance for fracking areas, including provisions for setbacks. These provisions provided local communities with a blueprint for local development. As such, they can be viewed as a solution to the problem of uncoordinated planning, as well as a solution to the problem of communities that may not have enough resources to come up with a rational comprehensive plan for shale development.

These provisions were challenged as an unjustifiable usurpation of community autonomy, and the Pennsylvania Supreme Court recently overturned the zoning provisions of Act 13. Thus, the extent to which the state can facilitate economic development is in question.

Regardless of the resolution of the legal dispute over authority to preclude local bans on fracking, it should be clear that communities have substantial autonomy even without authority to ban fracking. Communities cannot ban shale; however, they can regulate. The model zoning ordinance may also free up communities from conflict with gas companies. In some ways, Act 13 is a response to these problems of zoning. There is a larger literature that suggests communities can be influenced by business interests. In sociology and planning, this is often referred to as the growth machine: communities are often overwhelmed by business interests (Molotch 1976). There is also a literature of the structural dependence of the state on capital (Przeworski and Wallerstein 1988). This has led to effort to coordinate communities in the planning process or for state-level management of economic growth (Lubell et al. 2005; Feiock et al. 2008). State-level coordination on zoning is based on the realization that communities are not often able to coordinate on their response to development. The presumption is that state coordination will increase the autonomy of local governments to regulate in areas where they have capacity, as such regulation seeks to level the playing field between business and government.

New York, while nominally preserving local autonomy to ban fracking, undermines local autonomy with the moratorium. Local governments may be able to ban fracking, but they cannot promote it. However, if the ban is removed and the communities retain ability to ban fracking, then they would have substantial autonomy. Thus, there is a potential for substantial community autonomy in New York provided the moratorium is lifted, although until then, New York regulations undermine self-determination of communities.

Table 1 Comparison of Pennsylvania and New York responses

Dimensions	Pennsylvania	New York
Responsiveness to wealth	Excellent: Act 13 promotes fracking and maximizes the value of leases	Poor: Resources are wasted on leasing land without opportunities for production, although as shale remains in the ground, the value is not dissipated
Economic externalities	Excellent to fair: the model zoning provisions are a response to a local collective action problem, and there are few discernible economic externalities associated with fracking	Excellent for groundwater (no fracking issue), poor in terms of promoting alternatives to coal
State revenue	Excellent: the state includes provisions for impact of shale and includes guidance to ensure that the revenue is used to resolve economic externalities	Poor in the short run (government revenue precluded), although neutral in the long run, as governments may profit from shale production in the future through taxes and fees
Local autonomy	Excellent to fair: there is substantial autonomy for communities to zone, although they cannot ban fracking outright, and the zoning provisions free up communities to focus on other regulations	Poor under moratorium (no ability to allow fracking), excellent under lifted moratorium and home rule (authority to ban or allow fracking)

The comparison of the responses of each state on these dimensions is summarized in Table 1. On each dimension, the regulations in Pennsylvania appear to be efficient or at least include provisions that suggest the legislation is efficient. In contrast, the regulations in New York, at least in the short run, appear to be inefficient. Of course, the shale gas is not going anywhere in New York under the moratorium, and so it is important not to criticize the state's response too much, although it also remains fairly clear that the regulatory delay has important costs in the short term.

Explaining Regulatory Response

Once we understand the consequences of institutions, it is important to understand why they change. This section considers several common mechanisms of institutional change that can explain why Pennsylvania responded with efficient regulations and New York responded with costly delays.

One important explanation for variation in institutions is geography (Diamond 2005). Geography may explain why certain institutions are chosen. For example, differences in settler mortality rates are hypothesized to explain differences in political institutions, with higher settler mortality rates contributing to lower-quality institutions (Acemoglu et al. 2002). These perspectives suggest that geographic factors may lead to the emergence of more efficient institutions.

Another general explanation for variation in institutions is change in relative prices. There is a long-standing hypothesis in that property rights will change in response to relative prices such that institutions will change when it is efficient for them to change (Demsetz 1967; Barzel 1989). This perspective suggests that variation in shale policies may reflect the gains from shale extraction, with areas where there is more to gain from shale extraction more likely to adopt efficient regulations.

The structure of institutions may also influence regulations. The literature on path dependence stresses how institutional arrangements may induce different preferences (Pierson 2000; Acemoglu and Robinson 2006). One of the most important aspects of institutional variation in shale is with the underlying structure of ownership. For example, in Europe, the state typically owns mineral rights. In contrast, in the United States, individuals typically own mineral rights. Fee simple property rights are defined as those existing from above the surface but also below the ground (Ellickson 1993). The fee simple property system ensures that there will be a constituency—the mineral owners—with a direct interest in allowing fracking. In contrast, lease owners do not realize gains from shale directly, and so landowners in Europe have less of a direct interest in fracking. This perspective suggests that shale policies will be influenced by the presence or absence of private property rights to mineral land.

Although geography, relative prices, and institutions often explain the path of institutional change, each seems implausible in the case of state shale regulation. Despite some differences in shale, geography is similar enough to rule that out as an explanation for differences in shale production. Relative prices can also be ruled out, as there are tremendous gains from shale extraction in both states. Property rights may explain variation between the United States (where shale production has increased rapidly, and there are well-defined property rights to shale) and Europe (where shale production is almost nonexistent and the state often retains mineral rights). However, since Pennsylvania and New York have essentially the same underlying property institutions, we do not have to worry about differences in a direct interest in shale as an explanation and can rule it out as an explanation for institutional change.

This leaves politics and group conflict as potential explanations for institutional change. Political theories of property rights focus on political interests, political ideology, and political institutions as explanations for variation in institutions (Knight and Sened 1995; Sened 1997). These theories suggest that variation in regulations governing shale will reflect political considerations.

Political theories of institutional change often emphasize the interests of key political actors as a reason for institutional change. These theories suggest that institutions will reflect the interests of political officeholders and perhaps government bureaucrats (Riker and Sened 1991; Sened 1997). Act 13 was presided over by a Republican, Tom Corbett, who was elected in 2010. In New York, the moratorium emerged and has been continued by democrats. It is thus plausible that the interests of political elites explain variation in shale policies.

Public opinion also provides a plausible explanation for differences in policies. In democracies with winner-takes-all political districts, one obvious explanation for differences in public policies is the interest of the average voter (Mayhew 1974; Krehbiel 2010). One way to gauge the preferences of the average voter regarding fracking is through public opinion surveys. Such surveys are of limited value in a national sample, as many have limited knowledge of fracking in the United States as a whole, although there is much more awareness of fracking in states such as Pennsylvania and New York (Boudet et al. 2014). In Pennsylvania, polls suggest substantial uncertainty and no clear overwhelming odds in support of fracking, but a majority support fracking (Kriesky et al. 2013). There is clearly more of a divide in New York over fracking. These differences in perceptions of the benefits and costs of fracking are a plausible explanation for differences in public policy choice.

Group conflict is also a potentially important mechanism of institutional change (Knight 1992; Acemoglu 2003). These theories suggest that regulations and institutions will reflect the interests of powerful groups.

In the case of shale, there are several groups to consider. One is landowners. Landowners have an interest in supporting fracking; that maximizes the value of their lease rights. But they may also have signed poor leases. One possibility is that landowners who signed leases may now want regulatory delay. The landowners may actually benefit, especially the ones who signed leases on poor terms. In NY, leases began to be signed around 2000. Regulatory delays have led companies like Chesapeake to ask the courts to allow them to continue to hold leases without drilling (companies have to drill in order to maintain the lease). However, the author could find no evidence of landowners' associations or landowners who signed leases opposing a fracking. In addition, landowners' associations in Pennsylvania have generally been supportive of fracking, in particular since they often are able to bargain collectively for generous lease terms and inclusion of provisions that are favorable to surface owners in terms of liability for damages during the fracking process.

There is also a large literature on the importance of protest movement in understanding public policies (Tilly et al. 2001). There is no shortage of anti-fracking groups in New York. One activist group is "New Yorkers Against Fracking." A large number of municipalities have banned shale production, and local collective action appears to explain variation in these policies (Arnold and Holahan 2014). It certainly seems plausible that differences in the structure of protest activities may explain variation in public policies, although separating the impact of protest groups from political ideology of the public is both challenging and beyond the scope of this essay.

Political institutions are often used to explain variation in institutional change. However, these studies typically focus on the local level, such as with referendum and other policies (Feiock 2004; Lubell et al. 2005; Feiock et al. 2008). There may be variation in state-level political institutions that explain variation in the regulatory response, although it is more likely to find variation in political institutions of local government that explain whether a community bans or allows fracking (Table 2).

Table 2 Plausibility of alternative explanations

Theoretical mechanism	Plausible?	Rationale
Geography	No	Geographic similarity in Pennsylvania and New York can rule out geography as an explanation
Relative prices	No	There are similar gains from shale production in both states
Property institutions	No	These states have the same property system, leases signed rapidly in both, and landowners in both had an interest in shale production
Political interests	Yes	Variation in political control of the state may explain variation in support for fracking in Pennsylvania and the moratorium in New York
Political ideology	Yes	Public opinion polls show more support for fracking, and more belief benefits outweigh costs, in Pennsylvania
Group conflict	Yes	There are active protest groups in both states, although there appear to have been more local bans in New York, which suggests more protest activities
Political institutions	Unclear	There does not appear to be enough variation in political institutions at the state level to understand variation in fracking policies, although variation in local policies may be understood in terms of political institutions

Federalism and Fracking

The discussion so far suggests that there has been substantial variation in state responses to fracking. Is this an example of federalism gone awry? Or is the variation in fracking regulation an example of how federalism is supposed to work?

The case for federalism is well known, although it is useful to review it in light of many studies suggesting that regulatory variation with shale illustrates a flaw with federalism. One rationale is that federalism contributes to the emergence of market institutions while also improving the efficiency of public goods provision (Weingast 1995; Weingast 1997). Another perspective focuses on federalism in a more general sense, viewing it as an example of polycentric governance. Ostrom (1990, 2005) clarified the importance of polycentric governance of natural resources, suggesting that such governance is appropriate when it comes to natural resources since communities are often better able to understand and manage environmental issues than higher levels of government.

These perspectives suggest that there are theoretical benefits of decentralized regulation of shale. Contrast the logic above with a recent report issued by *Resource for the Future*, which is organized around the idea that

> Heterogeneity in and of itself is not good or bad. A main part of government's job is to internalize externalities, such as pollution. If the heterogeneity we observed reflects different conditions across states that lead to different levels of environmental risks, then that heterogeneity is a good thing. On the other hand, if the heterogeneity does not depend

> on environmental risks but is, perhaps, more dependent on politics, regulatory capture, economic concerns about jobs, or simply historical evolution or unexamined assumptions, we might question whether this heterogeneity is justified. Indeed, even if a state's regulations perfectly internalized in state externalities, these regulations may affect the environment in neighboring or downstream states. (Richardson et al. 2013)

The report concludes that there is no evidence of capture or a clear explanation for heterogeneity. Since they cannot identify differences in environmental externalities as a source of variation in state policies, the report concludes that heterogeneity of the response is viewed as somewhat troubling. Others criticize that fracking is exempted from the federal Safe Water Drinking Act (Warner and Shapiro 2013).

One of the weaknesses with these approaches is that they do not consider many reasons for federalism beyond the environmental coordination. As Spence (2013) recently argues, there are major opportunity costs from a moratorium on fracking, including economic benefits. As explained below, variation in political preferences, including perceptions about the importance of economic benefits from shale production, is an important reason why regulatory federalism is appropriate for shale, rather than evidence of a need for greater coordination by the central government.

Heterogeneity of Preferences

Federalism provides opportunities for public policies to account for diverse preferences. Public opinion surveys suggest that there are important differences in Pennsylvania and New York regarding shale. In New York, there are a majority that oppose, and in Pennsylvania, there are a majority that support. There is more support among Republicans for fracking. Independents tend to be split. In light of these ideological differences, it is unclear why it would be desirable to impose a common standard. Differences in state responses to fracking, to the extent they reflect underlying heterogeneity of preferences (including heterogeneity of economic conditions, which also factors into support and opposition to fracking), suggest that we should expect variation in policies governing shale and that such variation is part of how federalism is supposed to work.

Heterogeneity of Geography

A second general reason for federalism of fracking is variation in geography in the United States. Pennsylvania and New York are similar in that they each have fracking near groundwater (or potentially near groundwater). However, it is clear that the geography in Texas differs, where groundwater contamination is less of an issue. In Texas, it is more feasible to inject wastewater into EPA-sanctioned deep injection wells. In Pennsylvania, this is not so much of an option. This variation in geography suggests that federalism of fracking, whereby states can design different rules, is appropriate.

Experimentation

One of the rationales for federalism is that states are laboratories for policy experimentation. It was for this reason the states are often described as laboratories of democracy. States, by experimentation, can then be mimicked, allowing policies to diffuse through a decentralized process (Shipan and Volden 2006).

For some, uncertainty implies the importance of a national policy regarding fracking. However, uncertainty can also be a reason for decentralized governance of shale. Pennsylvania is an experiment with fracking; this will certainly be important information to New York. With a national policy similar to New York, there would be little to study regarding fracking. In this sense, federalism ensures that some states can be the first to experiment, and then other states can follow suit, modifying and adapting policy based on experience.

Local Administrative Capacity

It would make little sense to decentralize authority over fracking if the local units do not have the capacity to regulate shale production. For example, shale requires checking wells to see if there are any violations. In response to increasing shale permits and shale activity, the Pennsylvania State Department of Environmental Protection quadrupled the number of regulators assigned to shale. There are few reasons to doubt that New York would have the capacity to regulate as well. And if there is any doubt about the states, the EPA is conducting a massive study of fracking. As the national government provides oversight, it actually reduces the burden on local units.

Extent of Coordination Problems

Coordination problems are an important reason for the federal government. One of the challenges under the Articles of Confederation was lack of coordination on trade policy, as the states enacted tariffs against one another. The major change was to provide Congress with the authority to regulate commerce, including regulation of waterways.

It is not clear that there are coordination problems arising from different regulatory regimes at the state level. The various federal regulations of water are used to make a case for federal regulation of shale. However, the coordination problem lies within states; there are potentially many local zoning regulations. Act 13 is an example how to resolve the coordination issue. Groundwater may cross boundaries, but for the most part, shale production is an issue with local relevance.

Rabe (2014) suggests that there are few interstate compacts for different issues. This is true, but the fact remains that so much of shale is local, and so many of the impacts are local. There may not be a need for such compacts. Indeed, shale production seems to have produced few externalities, which suggests that more interstate regulation may be unnecessary.

The Price of Carbon

Shale is a fossil fuel, and so it is not a long-run solution to energy problems. There is a case to be made it is cleaner than other fossil fuels, but it still remains that there are unresolved externalities with fossil fuels, both conventional and unconventional. The question, in terms of the design of political institutions, is whether decentralized governance of shale production increases or decreases the chances of an appropriate price on carbon. States may not tax shale appropriately compared to the federal government. However, it seems that neither state governments nor Congress are willing to price carbon appropriately, and so it is likely that shale production, if it is produced at all, will be taxed at levels that do not fully internalize externalities associated with greenhouse gas production.

Conclusion

The shale gas boom provides an important example how political jurisdictions respond to new economic opportunities. In Pennsylvania, the response to fracking appears to be efficient, and in New York, it appears inefficient. The similarities of these states in terms of geography, relative prices, and institutions, and differences in politics, suggest it is politics that explains variation in these policies. The case of shale also suggests that there are virtues to a "fracking federation." Rather than a sign of regulatory incoherence, the radically different responses to fracking are examples of how federalism is supposed to work. After all, the constitution does not impose a standard of efficiency, or remedying environmental externalities, upon states, and so variation in the response to fracking—whether efficient or inefficient, or environmentally conscious or not—is to be expected.

Some have argued that there is no need to rush to develop shale gas since property rights are secure (Goldstein et al. 2013). However, there has also been a very strong and seemingly effective regulatory response to shale (Bloomberg and Krupp 2014; Krupp 2014). The question now seems to be, "Why wait?" This chapter suggests there are indeed few reasons to wait, although it also suggests that political considerations, rather than economic ones, will continue to constrain development of shale.

References

Acemoglu D (2003) Why not a political Coase theorem? Social conflict, commitment, and politics. J Comp Econ 31:620–652

Acemoglu D, Robinson JA (2006) De facto political power and institutional persistence. Am Econ Rev 96:325–330

Acemoglu D, Johnson S, Robinson JA (2002) Reversal of fortune: geography and institutions in the making of the modern world income distribution. Q J Econ 117:1231–1294

Arnold G, Holahan R (2014) The Federalism of Fracking: how the locus of policy-making authority affects civic engagement. Publius J Federal 44:344–368

Banerjee N (2012) Clouded readings of EPA study of Dimock water, featured in "Gasland." Los Angeles Times

Barzel Y (1989) Economic analysis of property rights. Cambridge University Press, New York

Bloomberg MR, Krupp F (2014) The right way to develop shale gas. New York Times

Boudet H, Clarke C, Bugden D et al (2014) "Fracking" controversy and communication: using national survey data to understand public perceptions of hydraulic fracturing. Energy Policy 65:57–67

Cathles LM III, Brown L, Taam M, Hunter A (2012) A commentary on "The greenhouse-gas footprint of natural gas in shale formations" by RW Howarth, R. Santoro, and Anthony Ingraffea. Clim Change 113:525–535

Clay KB (1999) Property rights and Institutions: congress and the California Land Act 1851. J Econ Hist 59:122–142

Clay K, Wright G (2005) Order without law? Property rights during the California gold rush. Explor Econ Hist 42:155–183

Coase R (1960) The problem of social cost. J Law Econ 3:1–44

Colaneri K (2014) Pa. towns with no zoning rules unlikely to limit gas drilling. StateImpact

Collier P, Hoeffler A (2005) Resource rents, governance, and conflict. J Conflict Resolut 49: 625–633

Cusick M (2014) Did Pennsylvania's highest court unravel environmental protections for oil and gas? StateImpact

Demsetz H (1967) Toward a theory of property rights. Am Econ Rev 57:347–359

Detrow S (2012) Estimate: impact fee would generate $219 million In 2012. StateImpact

Diamond J (2005) Collapse: how societies choose to fail or succeed. Viking, New York

Drajem M (2013) EPA official links fracking and drinking water issues in Dimock, Pa. The Washington Post

Ellickson RC (1993) Property in land. Yale Law J 102:1315–1400

Feiock RC (2004) Politics, institutions and local land-use regulation. Urban Stud 41:363–375

Feiock RC, Tavares AF, Lubell M (2008) Policy instrument choices for growth management and land use regulation. Policy Stud J 36:461–480

Fitzgerald T (2013) Frackonomics: some economics of hydraulic fracturing. Frackonomics Some Econ Hydraulic Fract Case Western Law Rev 63:1337–1362

Goldberg E, Wibbels E, Mvukiyehe E (2008) Lessons from strange cases democracy, development, and the resource curse in the US States. Comp Polit Stud 41:477–514

Goldstein BD, Bjerke EF, Kriesky J (2013) Challenges of unconventional shale gas development: so what's the rush. Notre dame J Law Ethics Public Policy 27:149–88

Governor's Marcellus Shale Advisory Commission (2011) Governor's Marcellus Shale Advisory Commission report. Harrisburg

Grossman PZ (2013) US energy policy and the pursuit of failure. Cambridge University Press, Cambridge

Hakim D (2013) New York Governor puts off decision on drilling. New York Times

Hills Jr RM (2014) Hydrofracking and home rule: defending and defining an anti-preemption canon of statutory construction in New York. Albany Law Rev 77: 648–72

Holahan R, Arnold G (2013) An institutional theory of hydraulic fracturing policy. Ecol Econ 94:127–134
Hopey D (2014) Corbett administration asks court to reconsider Act 13 decision. Pittsburgh Post-Gazette
Howarth RW, Santoro R, Ingraffea A (2011) Methane and the greenhouse-gas footprint of natural gas from shale formations. Clim Change 106:679–690
Howarth RW, Santoro R, Ingraffea A (2012) Venting and leaking of methane from shale gas development: response to Cathles et al. Clim Change 113:537–549
Knight J (1992) Institutions and social conflict. Cambridge University Press, New York
Knight J, North DC (1997) Explaining the complexity of institutional change. The political economy of property rights, Institutional Change and Credibility in the Reform of Centrally Planned Economies. Cambridge University Press, New York, pp 349–54
Knight J, Sened I (1995) Explaining social institutions. University of Michigan Press, Ann Arbor
Krehbiel K (2010) Pivotal politics: a theory of US lawmaking. University of Chicago Press, New York
Kriesky J, Goldstein BD, Zell K, Beach S (2013) Differing opinions about natural gas drilling in two adjacent counties with different levels of drilling activity. Energy Policy 58:228–236
Krupp F (2014) Don't just rill, baby – drill carefully how to making fracking safer for the environment. Foreign Affairs
Libecap GD (1989) Contracting for property rights. Cambridge University Press, New York
Libecap GD, Smith JL (1999) The self-enforcing provisions of oil and gas unit operating agreements: theory and evidence. J Law Econ Org 15:526–548
Libecap GD, Smith JL (2002) The economic evolution of petroleum property rights in the United States. J Legal Stud 31:S589–S608
Libecap GD, Wiggins SN (1985) The influence of private contractual failure on regulation: the case of oil field unitization. J Polit Econ 93:690–714
Lubell M, Feiock RC, Ramirez E (2005) Political institutions and conservation by local governments. Urban Aff Rev 40:706–729
Mayhew DR (1974) Congress: the electoral connection. Yale University Press, New Haven
McKinley J (2013a) Still undecided on fracking, Cuomo won't press for health study's release. New York Times
McKinley J (2013b) Fracking fight focuses on a New York Town's Ban. New York Times
McNulty T (2011) Mayor to let drilling referendum die. Pittsburgh Post-Gazette
Mittal S, Rakove JN, Weingast BR (2011) The constitutional choices of 1787 and their consequences. Founding choices: American Economic Policy in the 1790s. University of Chicago Press, Chicago, pp 25–56
Molotch H (1976) The city as a growth machine: toward a political economy of place. Am J Sociol 82:309–332
Muehlenbachs L, Spiller E, Timmins C (2014) The housing market impacts of shale gas development. National Bureau of Economic Research
Murtazashvili I (2013) The political economy of the American frontier. Cambridge University Press, New York
North DC (1981) Structure and change in economic history. W. W. Norton & Company, New York
North DC (1990) Institutions, institutional change and economic performance. Cambridge University Press, New York
North DC, Rutten A (1987) The Northwest Ordinance in historical perspective. Essays on the Economy of the Old Northwest, pp 19–31
North DC, Thomas RP (1973) The rise of the Western world: a new economic history. Cambridge University Press, New York
Olmstead SM, Muehlenbachs LA, Shih J-S et al (2013) Shale gas development impacts on surface water quality in Pennsylvania. Proc Natl Acad Sci 110:4962–4967
Ostrom E (1990) Governing the commons: the evolution of institutions for collective action. Cambridge University Press, New York
Ostrom E (2005) Understanding institutional diversity. Princeton University Press, Princeton

Pierson P (2000) Increasing returns, path dependence, and the study of politics. Am Polit Sci Rev 94:251–267
Pifer RH (2010a) What a short, strange trip it's been: moving forward after five years of Marcellus Shale development. U Pitt L Rev 72:615
Pifer RH (2010b) Drake meets Marcellus: a review of Pennsylvania case law upon the sesquicentennial of the United States Oil and Gas Industry. Tex J Oil Gas Energy L 6:47
Przeworski A, Wallerstein M (1988) Structural dependence of the state on capital. Am Polit Sci Rev 82:11–29
Rabe BG (2014) Shale play politics: the intergovernmental odyssey of American shale governance. Environ Sci Technol 48(15):8369–8375
Rabe BG, Borick CP (2012) Gas drillers' new Wild West. Philadelphia Inquirer
Revkin AC (2013) Is "Wild West" era for gas drilling coming to an end? New York Times
Richardson N, Gottlieb M, Krupnick A, Wiseman H (2013) The State of State Shale Gas Regulation. Resources for the Future Report
Riker WH, Sened I (1991) A political theory of the origin of property rights: airport slots. Am J Polit Sci 35:951–969
Riker WH, Weimer DL (1995) The political economy of transformation: liberalization and property rights. Modern political economy: old topics, new directions. Cambridge University Press, New York, pp 80–107
Ross ML (1999) The political economy of the resource curse. World Polit 51:297–322
Ross ML (2001) Does oil hinder democracy? World Polit 53:325–361
Schmidt CW (2011) Blind rush? Shale gas boom proceeds amid human health questions. Environ Health Perspect 119:a348
Sened I (1997) The political institution of private property. Cambridge University Press, New York
Shipan CR, Volden C (2006) Bottom-up Federalism: the diffusion of antismoking policies from US cities to states. Am J Polit Sci 50:825–843
Smydo J (2010) Pittsburgh moves ahead with controversial gas drilling Ban. Pittsburgh Post-Gazette
Smydo J (2011) Council OKs putting drilling ban on ballot. Pittsburgh Post-Gazette
Smydo J, Barcousky L (2011) County considering shale restrictions. Pittsburgh Post-Gazette
Spence DB (2013) Backyard politics, national policies: understanding the opportunity costs of national fracking bans. Yale J Reg 30:30A–475
Tilly C, McAdam D, Tarrow S (2001) Dynamics of contention. Cambridge University Press, New York
Umbeck J (1981) A theory of property rights: with application to the California gold rush. Iowa State University Press, Ames
Vidic RD, Brantley SL, Vandenbossche JM et al (2013) Impact of shale gas development on regional water quality. Science 340:1235009
Warner B, Shapiro J (2013) Fractured, fragmented federalism: a study in Fracking Regulatory Policy. Publius J Federalism
Weber JG (2012) The effects of a natural gas boom on employment and income in Colorado, Texas, and Wyoming. Energ Econ 34:1580–1588
Weber JG (2014) A decade of natural gas development: the makings of a resource Curse? Energy Res Econ 37:168–83
Weimer DL (1997) The political economy of property rights: institutional change and credibility in the reform of centrally planned economies. Cambridge University Press, New York
Weingast BR (1995) The economic role of political Institutions: market-preserving federalism and economic development. J Law Econ Org 11:1
Weingast BR (1997) The political foundations of democracy and the rule of law. Am Polit Sci Rev 91:245–263
Wiggins SN, Libecap GD (1985) Oil field unitization: contractual failure in the presence of imperfect information. Am Econ Rev 75:368–385
Wiseman H (2009) Untested waters: the rise of hydraulic fracturing in oil and gas production and the need to revisit regulation. Fordham Environ Law Rev 20:115

Sector Effects of Shale Gas Development

Alan Krupnick, Zhongmin Wang, and Yushuang Wang

Abstract This chapter reviews the impact of the shale gas revolution in the USA on the sectors of electricity generation, transportation, and manufacturing. Natural gas is substituting for other fuels, particularly coal, in electricity generation, resulting in lower CO_2 emissions from this sector. The use of natural gas in the transportation sector is currently negligible but is projected to increase with more investments in refueling infrastructure and better natural gas vehicle technologies. Petrochemical and other manufacturing industries in the USA and abroad have responded to lower natural gas prices by investing in US-located manufacturing projects.

Introduction

The shale gas revolution in the USA has significantly boosted US domestic natural gas production, which was previously in decline. US dry gas production increased by about 27.4 % from 18.05 Tcf in 2005 to 25 Tcf in 2012 (US Energy Information Administration [EIA] 2014 early release), largely because of the increasing production from shale gas and other unconventional sources. EIA predicts that this production will grow to 38 Tcf by 2040. This major shift in the supply of natural gas has driven down its price. The annual average Henry Hub natural gas spot price dropped by more than 50 %, from $8.86 per million Btu (mmBtu) in 2008 to $4.00/mmBtu in 2011, with a low of about $2.50/mmBtu in early 2012 and a return to about $4.50/mmBtu as of late March 2014.[1] These prices contrast to natural gas spot prices in Japan ranging from $13 to $15/mmBtu and in Europe of around $10/mmBtu. However, significant uncertainty is associated with the future price of natural gas, given uncertainty in demand, supply, and regulations, both directly on shale gas

[1] See EIA, "Henry Hub Gulf Coast Natural Gas Price", http://www.eia.gov/dnav/ng/hist/rngwhhdM.htm

A. Krupnick (✉) • Z. Wang • Y. Wang
Resources for the Future, 1616 P. St, NW, Washington, DC, USA
e-mail: krupnick@rff.org; wang@rff.org; ywang@rff.org

W.E. Hefley, Y. Wang (eds.), *Economics of Unconventional Shale Gas Development*, Natural Resource Management and Policy 45,
DOI 10.1007/978-3-319-11499-6_9

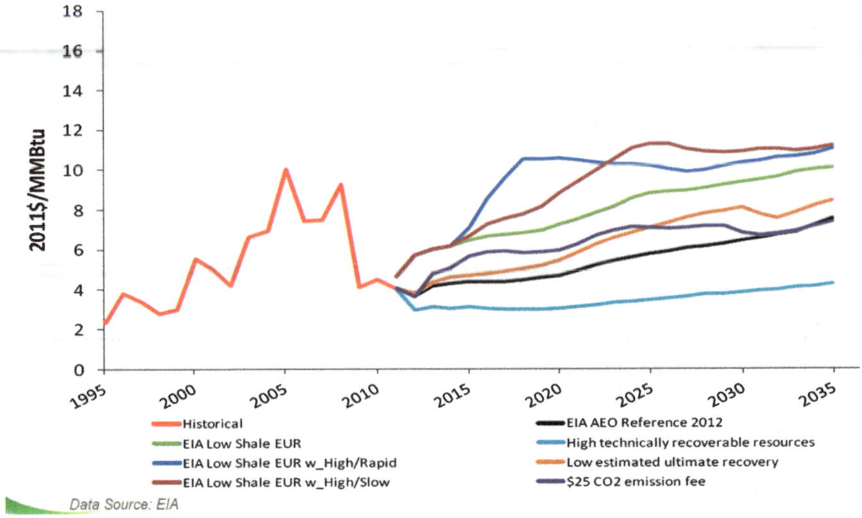

Fig. 1 Henry Hub natural gas spot price

extraction and through existing and potential climate policy.[2] Figure 1 shows the wide range of price forecasts EIA used to describe the future in its Annual Energy Outlook (AEO) 2012 forecast. A larger shale resource base assumption leads to a lower price projection and vice versa. Under a carbon pricing scenario, the natural gas price is projected to be higher relative to the reference case as a result of the demand shift to natural gas from more carbon-intensive fuels like coal.

The price decline has led to significant changes in the extent to which the USA uses natural gas in various energy-consuming sectors, including gas substitution for other fuels in the electricity, transportation, and industrial sectors. The purpose of this chapter is to document the effects in the USA that have already occurred, as well as those forecasted to occur in the future in the US market.[3]

In this chapter, we will first document the role that natural gas plays in the US economy. Following this, we provide a comprehensive review of the impacts of the shale gas boom on three end-use sectors—electricity, transportation, and manufacturing.

---

[2] Natural gas price volatility may be reduced as a result of changes in supply/demand balance and the geographic dispersion of shale plays, which would probably lower the importance of Gulf of Mexico as a source of gas supply (Lipschultz 2012).

[3] The US shale gas revolution has changed the energy landscape worldwide. For example, the shrinking demand for coal for power generation in the USA has resulted in an increase in US coal exports to the Europe, with coal replacing natural gas in European power sector. However, such effects on the energy market outside the USA are beyond the scope of this chapter.

Natural Gas in the US Economy

Natural gas has a unique place in the US economy, as it is a major fuel in the electricity, residential, and commercial sectors and a major feedstock for industry but has almost no role in the transportation sector. As shown in Fig. 2, natural gas represented 27 % of the primary energy consumed in the USA in 2012, accounting for 43 % of the energy supplied to industry, 75 % of energy supplied to residential and commercial heating and hot water, and 24 % of the fuels used to generate electric power. Natural gas supply to these sectors is split evenly across industrial, residential/commercial, and power sectors (about 32 % to each sector). In contrast, only 3 % of energy in the transportation sector is supplied from natural gas.

This chapter offers limited discussions of the residential and commercial sectors—largely because natural gas already has a 75 % share in these sectors. Further penetration would require more pipelines to be built to less densely populated areas, and in any case, turnover of heating and hot water systems would be slow. This is illustrated in Fig. 3, which shows that, while use of natural gas in power generation has escalated dramatically in recent years and industrial use of gas is enjoying a recent turnaround, the use of gas for commercial and residential heating has remained flat.

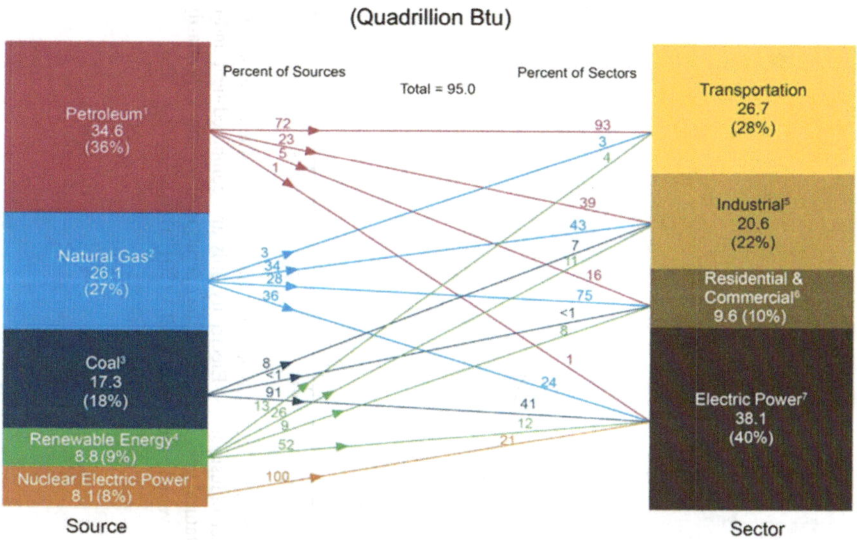

Fig. 2 Primary energy consumption by source and sector in 2012. (1) Does not include biofuels that have been blended with petroleum—biofuels are included in "Renewable Energy." (2) Excludes supplemental gaseous fuels. (3) Includes less than 0.1 quadrillion Btu of coal coke net imports. (4) Conventional hydroelectric power, geothermal, solar/photovoltaic, wind, and biomass. (5) Includes industrial combined-heat-and-power (CHP) and industrial electricity-only plants. (6) Includes commercial combined-heat-and-power (CHP) and commercial electricity-only plants. (7) Electricity-only and combined-heat-and-power (CHP) plants whose primary business is to sell electricity, or electricity and heat, to the public. Includes 0.2 quadrillion Btu of electricity net imports not shown under "Source" (Source: EIA)

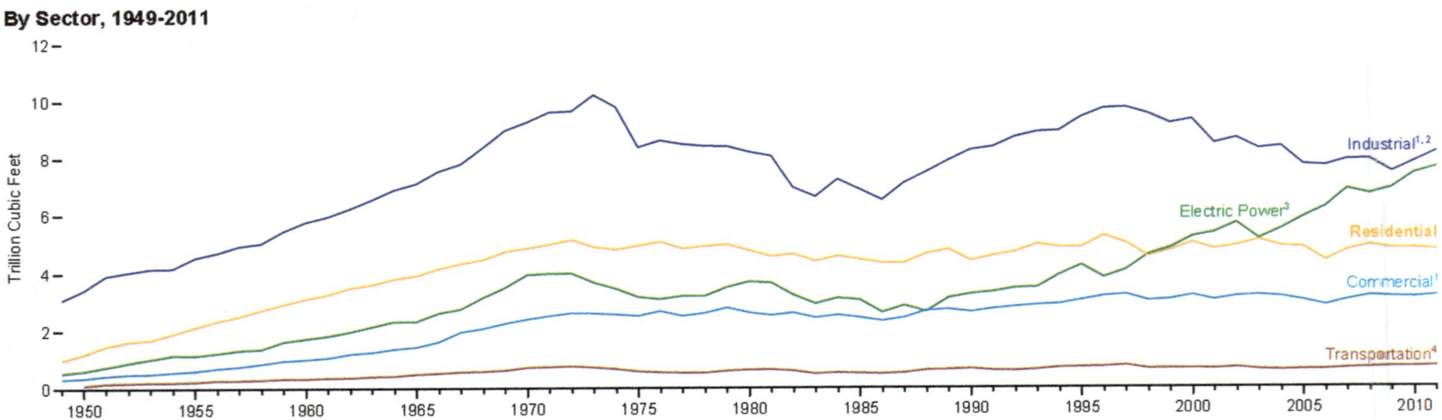

Fig. 3 Natural gas consumption by sector (Notes: (1) Includes combined-heat-and-power plants and a small number of electricity-only plants. (2) Includes lease and plant fuel and other industrial. (3) Electricity-only and combined-heat-and-power plants whose primary business is to sell electricity, or electricity and heat, to the public. (4) Natural gas consumed in the operation of pipelines (primarily in compressors) and as fuel in the delivery of natural gas to consumers; plus a small quantity used as vehicle fuel. Source: EIA (2012a)

The Electricity Sector

Lower natural gas prices are expected to drive more power plant operators to switch to natural gas from other fuel sources and therefore increase the share of natural gas in the power generation fuel mix. Similarly, and other things being equal, cheaper gas should decrease overall electricity prices and therefore increase the quantity of electricity demanded. At the same time, low natural gas prices can further disadvantage the economic case for nuclear power and renewables, while at the same time offer potential complements to intermittent renewable generation, thus potentially affecting these lower- or zero-carbon fuels.

Natural gas can be used by three generation technologies: natural gas combined cycle (NGCC) units, steam turbines, and gas turbines. Of these three technologies, NGCCs and steam turbines are usually used as base-load or intermediate-load units, while gas turbines are more likely to act as peaking units given their high flexibility (Massachusetts Institute of Technology 2009). In the near term, switching from other fuels to natural gas could be achieved by varying the capacity factors of different generating units—that is, running natural gas-powered generators more frequently to take advantage of the cheaper fuels. In the long term, the change in fuel prices would also affect business decision-making on new power plant investments and old plant retirements, thus changing the fuel mix of generating capacity.

Coal-to-Gas Switching

Although not as coal dominant as China's power generation sector, US electricity generation has also relied largely on coal. Recently, however, a low natural gas price has given natural gas-powered generation a competitive advantage over coal-powered generation, and there has been an evident trend to fuel switching—from coal and other fuel sources to natural gas—in the electricity fuel mix. From 2008 to 2011, the annual share of coal generation dropped from 48.2 % to 42.3 %, whereas the share of natural gas generation increased from 21.4 % to 24.8 %, and the share of renewable generation[4] increased from 9.2 % to 12.7 % (EIA 2012a). In fact, the share of coal in monthly generation dropped to the same level of natural gas generation for the first time in April 2012 (EIA 2012b).[5] Recent statistics indicate that the share of coal in annual generation hit a low level of 36 % as of August 2012 (Logan et al. 2012).

It is estimated that over 300 TWh of fuel switching from coal-fired to natural gas-fired electricity has occurred from 2008 to mid-2012 at the national level (Lee et al.

---

[4] Renewable generation includes traditional hydroelectric power as well as biomass, geothermal, solar, and wind generation.

[5] Note that this does not necessarily indicate an equal share of coal and natural gas in *annual* generation in 2012. This is because natural gas generation fluctuates a lot across different seasons of the year and is highly concentrated in the summer, when the peaking generators powered by natural gas are used to meet the high demand.

2012). For the USA, the potential to increase generation from existing gas-fired power plants is huge given its high natural gas generation capacity and relatively low utilization factors for gas-fired units in the years prior to large-scale shale gas production. As of 2011, natural gas represented a total summer generation capacity of 413 GW, which is the largest among all fuel sources (94 GW larger than coal capacity; EIA 2012a). A large number of these natural gas-fired plants were added from 1998 to 2003, and some of these were idle when gas prices were high before the shale gas boom. But as gas prices plummeted, these gas-fired plants became more favored by utility managers. For example, American Electric Power, one of the two biggest coal consumers in the USA, ran its gas plants at a 70 % capacity level in 2012, while its coal plants ran less than half of the time (Mufson 2012). The weighted average capacity factor of natural gas combined cycle plants in the PJM Interconnection's service territory more than doubled from 2008 to the first quarter of 2012 (Lee et al. 2012).

However, according to Macmillan et al. (2013), the coal-to-gas switching is subject to a theoretical ceiling at 613 TWh a year, which is about 13 % of US power generation in 2011, due to the location of excess gas-fired capacity, technological constraints, long-term coal contracts, and the transmission constraints for both gas and power. Assuming a gas price at $2.50–4 per million Btu, the 198 GW of open-cycle gas-fired plants are unlikely to compete with coal-fired plants due to their lower efficiency; lignite does not face the threat of being crowded out by natural gas given its extremely low cost (Macmillan et al. 2013).

The coal to gas competition for power generation was mostly concentrated in the eastern part of the country given the relatively higher coal price it is facing (Macmillan et al. 2013). As modeling results from Burtraw et al. (2012) show, more electricity is generated from natural gas under the *Cheap Gas*[6] than the *Expensive Gas* scenario. This trend is most pronounced in competitive regions.

In the long term, it is projected that roughly 30 GW of coal-fired plants, which comprise about 10 % of the total coal generation capacity, will be closed down by 2016, according to the announced retirement plans made by companies as of July 2012 (Celebl et al. 2012). Apart from the abundant supply of natural gas, the expected stricter regulation of air pollution from coal combustion also plays an important role in these anticipated closures. Many companies with older coal power plants have to decide between investing in environmental control facilities to ensure that their coal plants stay in operation versus putting that investment into new, cleaner gas-fired plants. Low gas price has made the latter choice more attractive to the industry, although the history of high gas price volatility acts to dampen the enthusiasm for natural gas.

[6]The Cheap Gas scenario reflects EIA's AEO 2011 projections of both electricity demand and natural gas supply, while Expensive Gas scenario uses the same AEO 2011 projection of electricity demand but substitutes EIA's projections of natural gas supply made in AEO 2009, which are much smaller. From AEO 2009 and AEO 2011, the unproved technically recoverable shale gas resource estimate increased by more than threefold from 267 Tcf to 827 Tcf (EIA 2012c). Relative to the Cheap Gas scenario, this scenario shows the effect on the electricity sector of lower natural gas supply and higher natural gas wellhead prices.

Mixed Effects on Renewables

In the short run, cheap gas would be unlikely to crowd out renewables since renewables have a lower variable cost as compared to almost all other resources due to the absence of fuel cost (Weiss et al. 2013). More importantly, renewables in many states command a mandated share of the generation mix. Thus, these limits constrain natural gas penetration. While in the long run, with the high up-front costs of renewables being considered, the improving economics of gas-fired plants will make them more competitive while competing with coal and renewable generations as options for new capacity additions (Weiss et al. 2013). Expectation of low gas prices in the near future would make it difficult for renewable energy project developers to sign power purchase agreements to finance their projects. Only those wind projects at favorable sites with government support can compete with natural gas generation on a levelized cost basis (Lee et al. 2012).

Meanwhile, there also exist potential synergistic opportunities between natural gas and renewables for electricity generation, ranging from "tightly coupled hybrid technologies[7]" to "more loosely coupled integrated system and market designs" (Lee et al. 2012, p. 3). A low natural gas price would make renewable generation more competitive by bringing down the cost of renewable-gas hybrid systems, in which intermittent renewable generation is backed up by flexible gas generation. Geographic overlaps between gas-producing regions and high wind energy potential regions indicate the possibility of jointly siting and developing wind and natural gas projects as well as the required transmission infrastructure (Lee et al. 2012). Hence, the long-run overall impact of cheaper natural gas on renewable generation depends on the characteristics of the power market, such as current capacity fuel mix, dispatching system, load characteristics, and relevant regulations.

While an expanded supply of natural gas could have mixed effects on renewables, modeling results (Burtraw et al. 2012) show that, by 2035, renewable generation is projected to be about 5 % lower in the *Cheap Gas* scenario compared to the *Expensive Gas* scenario. Such an overall "crowding-out" effect of cheap natural gas on renewables is consistent with the dominant industry view and supports the concerns of environmentalists that shale gas might hurt the market share of renewables.

Electricity Price, Demand, and GHG Emission Changes

Cheaper natural gas as a fuel source for generating electricity would potentially lower electricity prices. However, the moderate decline of average real electricity prices from 2008 to 2011 was driven by several factors, including the economic

---

[7] Examples include hybrid concentrating solar power (CSP) and natural gas-fired power generation systems, biogas and natural gas co-fired combined cycle gas turbines, natural gas-powered compressed air energy storage (CAES) to store non-peak renewable electricity generation for peak period usage, etc. (Lee et al. 2012).

downturn, energy efficiency improvements, and changed supply scenarios. After adjusting for inflation, average electricity prices decreased by about 2 % from 2008 to 2011, from 8.97¢/kWh to 8.81¢/kWh (measured in 2,005 dollars). Among the four end-use sectors, the residential sector faced the highest electricity prices in 2011, at an average of 11.8 cents per kWh. The residential sector was also the only end-use sector to see a slightly higher real electricity price in 2011 compared to 2008. The other three sectors (commercial, industrial, and transportation) experienced a drop in electricity prices during the time period, with a decrease of 5 %, 3 %, and 6 %, respectively, from 2008 to 2011 (EIA 2012a).

Looking into the future, modeling results (Burtraw et al. 2012) indicate that the forecasted high levels of supply of domestic natural gas will continue to substantially reduce retail electricity prices over the next 20 years. At the national level, the average electricity price in 2020 is projected to be about 5.7 % higher in the *Expensive Gas* case than the *Cheap Gas* case. Such effects are the most prominent in the competitive regions, where the projected average electricity price difference between the two cases in 2020 is 9.6 %, while the cost-of-service regions see a smaller price difference at about 3.6 %. Cheaper gas is expected to affect different customer groups differently. At the national level, the percentage difference is the largest for the industrial users (6.8 % in 2020), followed by commercial users (5.7 % in 2020), and residential users (4.6 % in 2020). Industrial users in competitive markets will probably enjoy the greatest benefit from cheaper electricity—the percentage difference in electricity price is projected to be as much as 14.5 % in 2020.

Given higher gas and electricity prices under the *Expensive Gas* scenario compared to the *Cheap Gas* scenario, consumers respond by using less electricity in the former scenario. Similar to the effects on electricity price, such effects are the most prominent for industrial users and competitive regions.

Generation fuel mix changes have also led to changes in greenhouse gas (GHG) emissions from electricity generation, which accounts for about 40 % of total carbon dioxide (CO_2) emissions in the USA. The CO_2 emissions from electricity generation, after fluctuating in the range of 2,346–2,413 million metric tons from 2005 to 2008, dropped by 9.09 % to 2,146 million metric tons in 2009, which was then followed by a slight increase to 2,258 million metric tons in 2010. Total fossil fuel-based CO_2 emissions from all end-use sectors followed a similar pattern, with a significant drop from 5,572 million metric tons of CO_2 in 2008 to 5,206 million metric tons in 2009 and a minimal increase from 2009 to 2010 (U.S. Environmental Protection Agency [EPA] 2012). As modeled by Burtraw et al. (2012), the increased use of natural gas in the *Cheap Gas* scenario reduces CO_2 emissions from electricity generation from 2,676 million tons in the *Expensive Gas* scenario to 2,579 million tons in 2035.

Effects on Grid Operation

The increasing use of natural gas as fuel sources for electric power generation also brought up concerns over the electricity reliability caused by gas pipeline capacity constraints and the interruptible pipeline service (Lee et al. 2012). Regions that heavily rely on gas generation are subject to the risk of natural gas and power price spikes in the winter caused by gas pipeline capacity constraint. With an increasing share of power coming from gas-fired power plants, several regulatory changes are being put in place to facilitate better coordination between power transmission operations and gas pipeline operations. FERC issued a final rule in November of 2013 "allowing interstate natural gas pipelines and electric transmission operators to share non-public operational information to promote the reliability and integrity of their systems" (FERC 2013). The New England ISO recently changed its day-ahead bids deadline from noon to 10 a.m. and market clearing deadline from 4 p.m. to 1:30 p.m. in order to better align with the day-ahead bidding schedule of the continental natural gas market.[8] These changes in grid operations are intended to address the reliability concerns caused by an increasing share of gas generation.

The Transportation Sector

Unlike the electricity sector, the US transportation sector currently sees only a small proportion of its energy use coming from natural gas. In 2012, natural gas accounted for 3 % of the total 26.7 quadrillion Btu of energy consumption for transportation, leaving petroleum the dominant fuel with a 93 % share (Fig. 2; EIA n.d.).[9] Nevertheless, low natural gas prices coupled with relatively high oil prices have made natural gas increasingly attractive as a fuel choice for transportation.

Figure 4 plots the ratio of oil prices to natural gas prices on a per energy unit basis, showing a rapidly rising ratio in the past few years that soared to 500 % as of late 2011. While other factors affect the fuel prices paid by consumers at the pump (e.g., fuel taxes,[10] infrastructure cost, supplier competitiveness, and delivery and storage cost), the price gaps between compressed natural gas (CNG) and other alternative fuels at the retail level have also widened recently. Low gas prices have undoubtedly quickened the trend of shifting from oil-based fuels to gas-based fuels in the transportation sector.

[8] http://isonewswire.com/updates/2013/5/22/spi-news-day-ahead-energy-market-timeline-changes-go-into-ef.html

[9] Natural gas consumed in the transportation sector includes both the use of natural gas to power natural gas pipeline transmission networks (2.8 % of total gas consumed in 2011) and natural gas used as vehicle fuel (0.1 % of total gas consumed in 2011) (Lee et al. 2012).

[10] "Currently, on a Federal level, [compressed natural gas] is taxed at the same rate as gasoline on an energy-equivalent basis ($0.18 per gasoline gallon equivalent, or 0.21 per diesel gallon equivalent), while [liquefied natural gas] is taxed at a higher effective rate than diesel fuel" (EIA 2012c, p. 38).

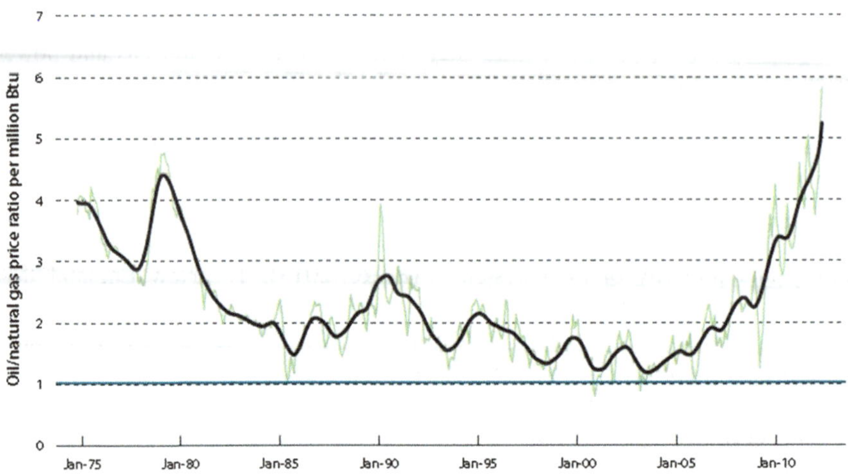

Source: Knittel (2012)

Fig. 4 Ratio of oil and natural gas prices per unit of energy (Source: Knittel (2012))

The Current Status of Natural Gas Use in Transportation

There are three basic ways for natural gas to replace oil for transportation use. First, natural gas can be converted to liquid fuels, such as methanol, ethanol, and diesel, through a gas-to-liquid (GTL) process, from which the liquid outputs can be burned in internal combustion engines with slight modifications. Second, CNG can be burned in light- and medium-duty natural gas vehicles (NGVs) or dual-fuel vehicles, which can run on either CNG or gasoline. Third, natural gas can be cooled and condensed into liquefied natural gas (LNG) that can then be used as a replacement for diesel, for use in heavy-duty trucks (as well as ships, barges, and railroads)

Market Penetration of Natural Gas Vehicles

NGVs have been a part of global vehicle fleets for decades, with an estimated 15.2 million on the road worldwide.[11] The USA currently ranks 17th globally in the number of NGVs on the road, behind countries such as Iran, Pakistan, Argentina, Brazil, India, Italy, and China.[12] In the past, NGV penetration in the USA was limited for the most part to small market niches: medium- to heavy-duty fleet vehicles, such as buses or trash trucks, and single-unit delivery truck fleets, such as those from FedEx,

[11] Natural Gas Vehicles for America, http://www.ngvc.org/about_ngv/index.html (accessed on 11/08/2013).

[12] Ibid.

UPS, AT&T (Taschler and Content 2011), and others. Although between 1999 and 2009 US domestic consumption of natural gas in the transportation sector tripled (Bryce 2011), overall use in the transportation sector remains very small; in 2010, natural gas powered less than 0.4 % of the nine million heavy-duty vehicles (HDVs) on the road and accounted for 0.3 % of total energy use by HDVs (EIA 2012c). By 2011, only about 0.05 % of vehicles in the USA were fueled by direct combustion of natural gas (Lee et al. 2012). Apart from the dedicated NGVs that run only on natural gas, the market offers bi-fuel vehicles, which have two separate fueling systems, enabling them to run on either natural gas or gasoline, and dual-fuels vehicles with fuel systems that run on natural gas and use diesel fuel for ignition assistance.[13] These technologies could serve as a bridge to the future with higher market penetration of NGV.

Public transit buses are the largest natural gas consumers in the transportation sector, with about 20 % of buses running on natural gas (C2ES 2012). Various public school districts have also converted their fleets to run on natural gas. For example, after taking part in a pilot alternative fuel vehicles project in the late 1980s, Tulsa Public School District in Tulsa, Oklahoma, now has a fleet of 190 CNG vehicles.

Interestingly, trash trucks have proven to be a major market niche for natural gas penetration, with estimates of 60 % of new trucks being powered by natural gas. It is reported that 15 cities and communities in the Northeast region had partially or totally shifted from diesel-powered refuse trucks to their analogies powered by natural gas as the end of 2012, which brought about $4.5–$6 million of fuel cost savings (Energy Vision 2013).

Considering light-duty vehicle (LDV) manufacturers, Honda recently launched its *2014 Civic Natural Gas* model with a combined fuel economy of 31 mpg, now available in 37 states; a Hong Kong-based company plans to build CNG/gasoline/electric hybrids in the USA; and Chrysler is gearing up to produce natural gas-fueled LDVs. Ford provided the market with the Transit Connect powered by CNG starting from 2011 and Vehicle Production Group came up with a natural gas-powered SUV model in 2012. As for truck engines, competition with industry leader Westport is growing from companies such as Emission Solutions, Inc. (ESI).[14] Chevrolet now offers bi-fuel Silver 2500HD that can seamlessly switch between natural gas and gasoline.

An absence of refueling infrastructure remains a significant impediment to broader penetration of NGVs in the USA, particularly outside of fleets that refuel in central locations. Trucks and buses often travel predictable routes and are stored in common areas, meaning that the infrastructure for a CNG fleet can be

---

[13] Natural Gas Vehicles for America, http://www.ngvc.org/about_ngv/index.html

[14] ESI has recently developed the natural gas-fueled Phoenix 7.6 L, a 300-horsepower rework of the heavy-duty Navistar MaxxforceDT diesel engine. Currently, ESI has plans to begin sales of the 375-horsepower Phoenix 9.3 L, project development on the T444E 7.3 L, and research and development on a 475-horsepower Phoenix 13 L in the third quarter of 2011 (Turner 2010).

concentrated in certain specified areas, so long as they are near gas pipelines, whereas the widespread use of CNG in passenger cars would require a much more extensive and costly refueling infrastructure (Alternative Fuels and Advanced Vehicles Data Center [AFDC] 2011). As of May 2012, 1,047 CNG fueling stations and 53 LNG fueling stations were in the USA, compared to 157,000 gasoline fueling stations nationwide in 2010. According to EIA (2012c), 53 % of the CNG stations and 57 % of the LNG stations are privately owned and not open to the public, and many of the public and private stations are concentrated in a few states like California. Accordingly, access to refueling infrastructure remains an obstacle in most parts of the USA. Part of the infrastructure challenge is the "chicken-and-egg" problem: vehicle users will not buy NGVs until they believe there are enough refueling stations, but motivation to build an NGV refueling infrastructure will be limited until a sufficient number of vehicle owners demand the fuel. Both the private and public sectors are working to address this issue, however, as described briefly later.

Gas-to-Diesel and Gasoline (GTLs) and Gas to Ethanol and Methanol

Through GTL technology, natural gas can be converted into diesel and gasoline, which can then be burned in traditional internal combustion engines. The conversion rate of current technology needs around 10 Mcf of natural gas as input for a barrel of oil-equivalent product. Assuming a $4/mcf gas price, this translates into a cost of $40/barrel oil equivalent (C2ES 2012). However, the high up-front capital cost of about $10 billion for a 100,000-bbl/day plant (Lipschultz 2012) remains a major issue for GTL projects. The use of GTL fuel in transportation remains limited since only a handful of GTL plants are operating commercially in Malaysia, South Africa, and Qatar today, and these plants are producing less than 1 % of global diesel demand. Nonetheless, the increasing availability of cheap gas has driven Sasol (a South African company) to announce plans to build the first GTL plant in the USA in Westlake, Louisiana with a total investment of $16–21 billion.[15] The project is currently in the FEED (front-end engineering design) stage and Sasol expects to reach final investment decision for the project in 2016.[16] However, Shell canceled plans for a GTL plant that had been expected to come online in 2019.

Natural gas can also be converted into methanol, ethanol, butanol, and DME, which can be mixed with gasoline in various fractions to create an alternative fuel. Common blends include E85 (85 % ethanol, 15 % gasoline) and M85 (85 % methanol, 15 % gasoline); these blends are currently usable by the 10 million flexible fuel vehicles on the road in the USA. Conversion kits are also available to allow standard internal combustion engine vehicles to run optimally on these blends.

---

[15] http://www.sasollouisianaprojects.com/page.php?page=projects (accessed on 05/29/2014).

[16] http://www.ogj.com/articles/2013/11/sasol-lets-contract-for-louisiana-gtl-plant.html

Forthcoming research (Fraas et al. 2013) indicates that, even incorporating the cost of a conversion kit, the wider use of E85 in passenger vehicles may make strong economic sense, given current fuel price differentials. This is based on estimates of the cost of producing ethanol (and eventually E85) from natural gas, using Celanese Corporation's "TCX" process. As more details become available about the costs of this process, the use of blended fuels may very well look increasingly promising to consumers and manufacturers.

Federal and State Efforts

The federal government has been trying to stimulate the use of natural gas in transportation through a series of subsidy programs. The Energy Tax Policy Act of 2005 (PL 109–58) provided an income tax credit for the purchase of a new, dedicated alternative fuel vehicle of up to 50 % of the incremental cost of the vehicle, plus an additional 30 % if the vehicle met certain tighter emission standards. These credits ranged from $2,500 to $32,000 depending on the size of the vehicle. However, the credit was effective only on purchases made after December 31, 2005, and expired on December 31, 2010.[17] In August 2009, the US Department of Energy (DOE) announced that funding for natural gas technologies and fueling stations would be included in a $300 million grant under the American Recovery and Reinvestment Act for state and local governments (PL 111–5). A more recent legislative effort was the House of Representatives 1,380 bill, the New Alternative Transportation to Give Americans Solutions (NAT GAS) Act in 2011.[18] This proposed legislation offers tax credits for new NGVs at the retail and manufacturing ends, commercial and residential refueling infrastructure, and the gas itself.[19] However, the NAT GAS Act was rejected by the Senate in March 2012. In early 2013, federal NGV tax incentives got passed as part of the American Taxpayer Relief Act of 2012 (HR 8; PL 112–240), which included "a 50 cent credit per gallon or gasoline gallon equivalent for the sale of natural gas as a motor fuel and a credit for 30 percent of the cost of installing new natural gas refueling equipment for up to $30,000."[20]

---

[17] PL 109–58 also provided for a tax credit of 50¢ per gasoline gallon equivalent of CNG or liquid gallon of LNG for the sale of CNG and LNG for use as a motor vehicle fuel. The credit began on October 1, 2006, and has recently expired. Note that this rebate (which is over twice the excise tax rate paid now) was to the seller, not the buyer. It is not clear if this could have been paid to the ultimate seller—in which case an owner of a trucking company could have qualified for the rebate—or to the wholesaler.

[18] Available at http://www.govtrack.us/congress/bill.xpd?bill=h112-1380

[19] Specifically, the NAT GAS Act offers (1) a tax credit for new NGV purchases, up to 80 % of the price differential, which translates to a maximum of $7,500 for LDVs and $64,000 for HDVs; (2) an infrastructure tax credit of 50 % of the cost of a new station, up to a maximum of $100,000; (3) an extension of the 50¢ per gallon fuel tax credit; (4) a $2,000 tax credit to home refueling units; and (5) a tax credit to NGV manufacturers. (Gray 2011).

[20] http://www.ngvamerica.org/gov_policy/fed_legislate.html (accessed on 05/19/2014).

At the federal level, in August 2011, EPA and the US Department of Transportation's (DOT's) National Highway Traffic Safety Administration adopted the first-ever program to reduce GHG emissions and improve fuel efficiency of medium-duty vehicles and HDVs, where NGVs and other alternative fuel vehicles were credited based on their GHG emission reduction potentials (EPA and DOT 2011). In March 2012, President Obama announced a new $1 billion National Community Deployment Challenge to "spur deployment of clean, advanced vehicles in communities around the country (White House 2012, p. 1)." This "fuel-neutral" proposal includes electrification, natural gas, and other alternative fuels. The program also seeks to develop up to five regional LNG corridors to increase NGV deployment (White House 2012). More recently, the president outlined a tax credit "for 50 percent of the incremental cost of a dedicated alternative-fuel truck for a five-year period" and committed financial support for a select number of natural gas vehicle deployment communities (White House 2013).

States and localities have also intervened. Due in part to air quality management district regulations, 65 % of all South Coast Air Basin transit buses are now fueled by natural gas. The San Pedro Bay Clean Air Action Plan, approved in late 2006, includes a program to replace all diesel trucks based in the ports of Los Angeles and Long Beach with clean alternatives, such as LNG-fueled vehicles (including LNG-fueled 18-wheelers), within 5 years (Port of Los Angeles and Port of Long Beach 2011). As of 2011, 879 natural gas-fueled trucks are in the Drayage Truck Registry, which represents 7 % of container trips in San Pedro Bay. Pennsylvania, a state with significant shale gas reserves, introduced a package of legislation aimed at providing $47.5 million in tax incentives, grants, and loans to promote investment in natural gas truck and bus fleets for municipalities and businesses.[21]

Regional efforts are also in place to address the chicken-and-egg problem by incentivizing or providing refueling infrastructure. Utah has been promoting the use of NGVs, including private automobiles, by working with a local gas utility to build the fueling infrastructure. Trailing only California and New York, Utah currently is one of the top states in terms of the number of CNG refueling stations, with 73 (AFDC 2011). In Colorado, the city of Grand Junction opened its first CNG refueling station in April 2011, completing a chain of CNG stations from California to Denver (Cianca 2011). Texas is building refueling stations between Dallas, San Antonio, and Houston under the Texas Clean Transportation Triangle strategic plan. Similar efforts are also under way in the western coast area (the Interstate Clean Transportation Corridor) and Pennsylvania (the Pennsylvania Clean Transportation Corridor) (EIA 2012c).

[21] The Marcellus Shale Coalition, a natural gas trade group in Pennsylvania, released a study in April 2011 to spearhead a campaign for 17 new refueling stations statewide and subsidies for a proposed 850 new natural gas HDVs for an estimated $208 million (Gladstein, Neandross and Associates 2011).

Private Efforts

Private corporations are playing an important role in promoting natural gas use for transportation, without government subsidies. The most important example is an effort spearheaded by Chesapeake Energy's $150 million commitment, collaborating with GE, Clean Fuels, and Pilot Flying J truck stops to develop 150 CNG and LNG refueling stations on Pilot Flying J footprints on US interstates (150 stations in all). GE will provide modular and standardized CNG compression stations, called "CNG In A Box™."[22, 23] Private–public efforts to reduce the cost of home CNG refueling stations from their current cost of $4,000 are also ongoing (Lipschultz 2012).

The Current Economics of NGVs Versus Gasoline- or Diesel-Fueled Vehicles

The future role of natural gas in the US transportation fuel mix depends on the attractiveness of NGVs, compared to their alternatives, to consumers and policymakers. In this section, we investigate the evidence for and against NGVs as a reasonable option to their closest alternatives in the USA, focusing primarily on: (1) LDVs running on CNG compared to conventional gasoline vehicles and electric hybrids and (2) heavy-duty trucks running on LNG compared to diesel trucks. Many of the comparisons are based on several original analyses, using data from the NEMS-RFF model, automobile manufacturers, and other key sources.

The results suggest that, under reasonable conditions, LNG heavy-duty trucks have attractive payback periods even without government subsidies. Infrastructure issues may be less challenging than commonly thought because the interstate trucking industry is moving increasingly from a long-haul route structure to a "hub and spoke" structure—a development that could facilitate more judicious placement of LNG refueling stations and therefore make use of LNG trucks more prevalent (Taylor et al. 2006).[24] Furthermore, as noted above, efforts by Shell and Chesapeake Energy to build LNG refueling infrastructure represent a very positive and subsidy-free development. CNG as a fuel for LDVs remains a tough sell without policies that price carbon or otherwise favor natural gas over oil.

---

[22] "GE and Chesapeake Energy Initiative Targets Natural Gas Fueling Infrastructure Development," *NGV Global News*, http://www.ngvglobal.com/ge-and-chesapeake-energy-initiative-targets-natural-gas-fueling-infrastructure-development-0309

[23] Chesapeake Energy Corporation, "Transform U.S. Transportation Fuels Market and Increase Demand for U.S. Natural Gas," http://www.chk.com/About/BusinessStrategy/Pages/Increase-Demand.aspx

[24] See http://scm.ncsu.edu/public/lessons/less031014.html for a discussion of this system for major retailers in the USA.

Light-Duty Vehicles

Table 1 displays the relative differences in characteristics and costs among Honda's 2011 model year NGV (the Civic GX Sedan), a comparably equipped Honda Civic Sedan (the LX-S automatic transmission), and the Civic Hybrid (CVT AT-PZEV). Without a subsidy (the appropriate way to compare the costs of vehicles from society's point of view), the NGV is more expensive than the hybrid, but substantially (32 %) more expensive than the gasoline version. Its maintenance and repair costs are also more expensive than those for the other vehicles, over 50 % more than the gasoline-powered version.[25] The fuel economy for the NGV is about the same as that of the gasoline alternative (and far lower than the hybrid).

Assuming a $1.50 gal of gasoline equivalent advantage for natural gas over gasoline, 7 years of annualization, and a 6 % interest rate, and without counting infrastructure cost or any subsidies, we found that a natural gas LDV is almost $200 more expensive annually than its gasoline-fueled counterpart. Infrastructure costs for the NGVs must be considered, however, under the assumption that individuals will not purchase such vehicles unless they have access to a home fueling unit and already have natural gas in their homes. These units cost $4,000 currently. We assume they last 10 years and amortize their costs at the same 6 % interest rate.[26]

Table 1 Salient differences between NGVs and alternatives

Characteristic	Civic natural gas	Civic gasoline	Civic electric gasoline hybrid
MSRP (comparably equipped)	$26,240	$19,905	$24,700
Subsidy (eliminated January 2011)	$4,000	0	0
5-year maintenance and repair	$3,321	$2,145	$2,340
Combined fuel economy (mpg)	28	29	41
Fuel capacity (gge)	7.8	13.2	12.3
Range (mile)	218	383	504
Cargo volume (ft^3)	6	12	10.4
Availability	50 states	50 states	50 states
Total costs/year differential (without infrastructure)	$200	–	$400
Total costs/year differential (with infrastructure)	$721	–	$400
With $2,000 infrastructure subsidy and $4,000 vehicle subsidy	($100)	–	$400

Note: *gge* gallons of gasoline equivalent
Source: Honda website: http://automobiles.honda.com/tools/compare/

[25] These estimates are taken from Honda's own website. A similar comparison (Goulding et al. 2011) uses information from a Kansas Gas Service website, which asserts that "Some fleet operators have reduced maintenance costs by as much as 40 percent by converting their vehicles to CNG" (http://www.oneok.com/en/KGS/CustomerCare/BusinessDevelopment/NaturalGasVehicles.aspx).

[26] Honda Corporation also notes (personal communication) that high water content in the natural gas and low compression by home refueling units raises risks of fuel fouling in CNG engines.

Adding this annual amount to the annual cost of an NGV raises its cost premium over a gasoline-fueled counterpart from $200 to $721.

From an individual's perspective, we need to consider the $4,000 subsidy for the investment cost (which ran out at the end of 2010, but may be reinstated by federal legislation currently under consideration), the $2,000 subsidy for home charging stations, and the annual cost of the loan (which we assume is for a 5-year period). After these adjustments, amortized costs are about $100 less than a gasoline vehicle. As noted above, however, NGVs have much lower range and less trunk space and, in almost all US locations, could not reliably be used for long-distance travel because home refueling would be impossible. It remains to be seen if these restrictions are worth more to consumers than $100 per year.

Heavy-Duty Trucks

The 2011 national average retail price of diesel fuel was $3.84/gal, and the average nationwide nominal retail price was $3.05 per diesel gallon equivalent (dge) for LNG and $2.32/dge for CNG, which indicates a price differential of about $0.80/dge for diesel-fueled HDVs (EIA 2012c).[27] In California, where truckers can fill up with LNG at several stations, LNG is $0.75 per diesel gallon equivalent cheaper than diesel for an independent trucker and $1/gal cheaper for a fleet vehicle.[28] Indeed, when oil prices were at their highest in 2008 and diesel was $4.75/gal, LNG was $2/gal cheaper than diesel, even though natural gas was priced relatively high at $11–$13/mcf of gas (EIA 2008).

Table 2 gives the major assumptions when we compare natural gas-powered heavy-duty trucks with their diesel-powered counterparts. According to estimates available online and provided in conversations with experts, the investment cost differential ranges from $70,000 to $100,000 (for early models) more than the price of a diesel truck of about $100,000.[29] Detailed information on vehicle prices puts the cost differential at $70,000 for a Westport compression-type LNG engine, with a newer technology relying on an 85 % LNG/15 % diesel fuel mix, selling for only $35,000–$40,000 above its diesel counterpart.[30] The price differential for a smaller version of the Class 8 truck (termed a "Baby 8") or a Class 7 truck (both using spark plug technology) is around $40,000.

[27] Irrespective of these price differentials, it is appropriate to consider any tax benefits for natural gas over diesel. Currently, no such benefits are available. Until the end of 2009, LNG sellers were eligible for a credit of 50¢/gallon from the federal government (and some state programs provide per gallon credits against excise taxes). It is likely that some of these benefits would have been passed on in lower fuel prices.

[28] Interview with Mitchell Pratt, Clean Energy Inc., November 17, 2009.

[29] Total Transportation Services recently purchased 22 additional Kenworth T800 LNG trucks to expand its fleet of 8 such trucks purchased six months before. This purchase suggests that fuel and maintenance costs are manageable (Kell-Holland 2009).

[30] Interview with Michael Gallagher, Cummins Westport, November 2009.

Table 2 Assumptions for comparing natural gas heavy-duty trucks with diesel trucks

Price differential between LNG and diesel	\$0.50/dge, \$1.00/dge, and \$1.50/dge
Investment cost differential	\$35,000, \$70,000, and \$100,000
Fuel economy	Diesel (Class 8): 5.1 mpg (2007)[a]; LNG: 4.6 DEG to 5.6 DEG[b]
Vehicle miles traveled	70,000 miles/year[c] to 125,000 miles/year[d]
Vehicle lifetime	15 years[e]
Interest rate	31 %,[f] 10 %, 5 %[g]

Note: *dge* diesel gallon equivalent

[a]FHWA (2008). This estimate was recently revised upwards to 6.0 mpg (FHWA 2009)

[b]Interview with Mitchell Pratt, Clean Energy Inc., November 17, 2009

[c]FHWA (2008)

[d]This is based on census data from 2002, which feature average vehicle miles traveled of about 90,000 miles per year and indicate that about one-third of the fleet drives 125,000 miles or more

[e]According to DOT, new combination trucks (Class 8) were purchased in 2007, with registrations in 2007 of 2.221 million combination trucks. Thus, new vehicles are 6.8 % of the fleet. Assuming this is an equilibrium situation, where truck retirements and purchases are equal, truck life averages 14.7 years. Industry analysts offer 18–20 years as a realistic average for truck life (FHWA 2008)

[f]This rate derives from actual market data showing that buyers demand a payback of investment costs through fuel savings within three to four years and that fuel savings during those first few years are discounted at 10 %

[g]Social discount rates used to evaluate public projects are often in the range of 3 % to 5 %. Although the substitution of NGVs for diesel vehicles is not a public project, it can confer major public benefits in terms of emissions reductions and energy security. Thus, we make calculations with this rate to illustrate the efficiency of LNG truck subsidies or mandates from society's perspective, assuming complete market failure. An interest rate of 10 % is added to reflect partial market failure

The payback period estimates based on the assumptions above are shown in Table 3. To get to payback periods of 2 years or less—what is commonly believed to be what industry is looking for before it makes investments—for an investment cost difference of \$70,000 and fuel economy of 5.1 miles per gallon equivalent, one needs fuel price differentials of around \$1.50 per gallon equivalent, rates of interest used to evaluate multiyear fuel savings benefits of 10 % or less, and vehicle miles traveled of around 125,000 per year. For lower-mileage trucks (90,000 miles per year), payback periods increase about a year. This finding indicates that the high-mileage part of the trucking fleet is most likely to be early adopters. Halving the fuel price differential more than doubles the payback period. Halving the investment cost differential more than halves the payback period (indicating the efficacy of rebates and subsidies). A 10 % increase in fuel economy of the LNG truck, other things being equal, leads to about a 10 % decrease in payback period at a fuel price differential of \$1.50, but this improvement leads to much greater payback period reductions when the price differential is smaller (i.e., less advantage to LNG over diesel). For instance, at a price differential of only \$0.75 per gallon equivalent, payback periods fall by about 20–25 %. These results indicate the sensitivity of payback periods to price fluctuations.

Table 3 Sensitivity of payback periods to assumptions

Vehicle cost differential:		$35,000			$70,000		
Fuel economy (mpg):		5.6	5.1	4.6	5.1		
Vehicle miles traveled:		70,000			125,000	90,000	70,000
Interest rate = 0.05	Fuel price diff. = $1.50	1.62	1.82	2.14	2.05	2.91	3.82
	Fuel price diff. = $0.75	3.04	3.82	5.54	4.33	6.29	8.52
	Fuel price diff. = $0.50	4.3	6.03	11.98	6.89	10.36	14.62
0.10	Fuel price diff. = $1.50	1.73	1.95	2.31	2.22	3.22	4.36
	Fuel price diff. = $0.75	3.39	4.36	6.74	5.03	7.9	11.96
	Fuel price diff. = $0.50	4.99	7.48	22.72	8.88	16.54	–
0.31	Fuel price diff. = $1.50	12.09	–	–	3.3	6.35	–
	Fuel price diff. = $0.75	–	–	–	–	–	–
	Fuel price diff. = $0.50	–	–	–	–	–	–

Other Considerations Affecting Cost

Even proponents of natural gas concede that NGVs face significant obstacles to capturing a major share of various market segments. Irrespective of vehicle type, observers have raised concerns regarding economics—NGVs cost more, although fuel costs are likely to be lower—as well as concerns about safety and the availability of refueling stations. There are also concerns about resale markets, which are an important part of the trucking industry and, if fueled by natural gas, require a denser refueling network than is likely to arise in the near term. In addition, for LDVs, cruising range, weight, and cabin space are subjects of concern. Because CNG has such a low energy density and is under pressure, fuel tanks are large and heavy compared to the other vehicle types. As Table 1 shows, cargo space is dramatically (50 %) lower than that of a gasoline vehicle, as is its range of only 218 miles, compared to 383 miles for the comparable gasoline vehicle and 504 miles for the hybrid.

Notably, the estimates above do not directly account for safety and infrastructure costs. There are arguments on both sides of the safety issue: proponents, for example, suggest that the need to contain high pressures and keep temperatures low requires extremely robust tanks and other equipment that may make natural gas trucks safer in an accident than their diesel counterparts. Opponents refer to concerns about LNG storage facilities and their explosive potential. An independent review of safety concerns (Hesterberg et al. 2009, p. 20) finds that diesel buses have a "significant fire and safety advantage over CNG vehicles [buses]." Whether these conclusions would hold for LNG versus diesel trucks is unclear. A government source[31] focusing on CNG versus LNG concludes that the latter is less corrosive but cannot take an odorant, so leaks could go undetected longer, requiring methane detectors. With

[31] See http://www.chebeague.org/fairwinds/risks.html, which is an excerpt from a report produced by the Federal Transit Administration's Clean Air Program, Sect. 3.3.4 Liquefied Natural Gas.

respect to LNG, the very cold temperatures required for storage mean that the storage systems require intensive monitoring for tank pressure and systems to vent the gas in an emergency. While the report indicates that rupturing of the tanks is extremely unlikely, it also says that any resulting fire will release 60 % more heat than from an "equivalent" gasoline tank rupture. Refueling NGVs also requires additional precautions, and the rapid change in temperature from refueling can stress vehicle materials and components. The industry's response to these points is basically that the fuel is safe if the proper procedures are followed.

In addition, even if the trucking industry had adequate refueling infrastructure for long-haul trucking, economic issues concerning lack of infrastructure appropriate to the truck resale market may remain. Trucks are sometimes taken out of the commercial trucking business and resold for use on farms and within cities after 6–8 years of use. Without adequate infrastructure in rural and urban areas, this market could fail, effectively limiting the useful life of these trucks, both from a private and a social perspective.

Future Projections of NGV Penetration

Looking ahead, expected future new vehicle cost differentials may be lower. First, NGVs have not yet benefited from economies of scale as gasoline and diesel vehicles have, so costs might decrease significantly if demand for NGVs increases. Second, stricter standards on diesel emissions, which took effect in 2010, may raise prices on diesel vehicles. Further, Phase II of the truck CAFE standards is slated to go into effect in 2016. These standards will affect diesel-fueled and alternate-fueled trucks alike, with adjustments made for CO_2 equivalent emissions. The effects of relative costs are unknown. In general, natural gas engine technologies are less mature than diesel and gasoline technologies, and it is uncertain which particular natural gas engine types will be most successful in the future and what their costs will be. But the relative immaturity of the natural gas engines makes it likely that the pace of innovation will be greater.

For several reasons, the recent fuel price gap could remain or widen in the future. Greater accessibility and technological advances in recovering shale gas could keep prices of LNG stable or even drive them lower, while prices for oil and, therefore, diesel fuel are determined on a world market. A recent presentation by IHS Global Insight (2010) shows that, over the long term, the ratio of oil to gas prices may rise to about three to one between now and 2030. However, natural gas prices have a history of instability, and CNG has, at times, been more expensive per gallon equivalent than its diesel counterpart.

In the AEO 2012, EIA runs a side case known as the *Heavy-Duty NGV Potential* case, in which natural gas refueling infrastructure is expanded (simply by assumption) and a gradual increase is allowed in the share of HDV owners "who would consider purchasing an NGV if justified by the fuel economics over a payback distribution with a weighted average of 3 years" (EIA 2012c, p. 39). In addition, an *HDV Reference* case was developed from the AEO 2012 *Reference* case, assuming

Table 4 Major projections under two heavy-duty NGV cases in AEO 2012

	2010	HDV reference case (2035)	Heavy-duty NGV potential case (2035)
Sales of new heavy-duty NGVs	860 (0.2 %)	26,000 (3 %)	275,000 (34 %)
Market share of heavy-duty NGVs	0.4 %	2.4 %	21.8 %
Natural gas demand in the HDV Sector	0.01 Tcf	0.1 Tcf	1.8 Tcf
Share of natural gas in total energy use by HDVs	0.2 %	1.6 %	32 %

Source: EIA (2012c)

that Class 3–6 vehicles use CNG and Class 7 and 8 vehicles use LNG. Table 4 summarizes the projected sales, market penetrations, and natural gas consumptions in the HDV sector in these two different scenarios in 2035. The wide gap between these two cases reflects a great uncertainty over the future prospect of NGVs. The higher consumption of natural gas in the *Heavy-Duty NGV Potential* case slightly pushes natural gas prices up, which results in lower gas consumption in other end-use sectors. The overall impact brings about a 5 % higher total US natural gas consumption compared with the *Reference* case (EIA 2012c). In AEO 2014, EIA (2014) added LNG as a fuel option for freight rail and domestic vessels and projected that the penetration of natural gas will reach 35 % for freight rail energy consumption and 2 % for domestic marine vessels by 2040. Natural gas use in LDVs, HDVs, locomotives, buses, and marine vessels is predicted to reach 863 trillion Btu by 2040, a 20-fold increase from 43 trillion Btu in 2012 (EIA 2014).

In summary, the economics of natural gas penetration into transportation suggests that this fuel deserves more attention. Honda's natural gas-fueled LDV needs investment and infrastructure subsidies at the level being discussed in Congress to compare favorably to its gasoline and hybrid counterparts. Under certain assumptions about fuel and vehicle price differentials, fuel economy, and vehicle miles traveled (such as being driven 125,000 miles per year), LNG-fueled heavy-duty trucks can return their added investment in 2 years, but generally, payback periods would be longer. Additionally, this somewhat optimistic assessment does not directly account for infrastructure and safety costs.

Nonetheless, a variety of developments are in play to make NGVs economical even without subsidies on the fuel or the vehicles. First, natural gas prices are projected to remain relatively low given vast new amounts of shale gas becoming available, even if demand increases greatly. Second, technological changes for NGVs are likely to be more rapid than those for conventionally fueled vehicles because the latter are more mature technologies. Third, if demand for NGVs does increase, economies of scale could further reduce prices. Fourth, diesel vehicles may become more cost disadvantaged in the future by a carbon policy combined with increasingly stringent air pollution regulations and tighter restrictions on fuel economy of gasoline and diesel vehicles (as well as natural gas vehicles). Fifth, technological advances in converting gas to ethanol, in particular, have the potential to replace oil without requiring as much infrastructure investment as CNG or LNG.

Indirect Effects

Paradoxically, low natural gas prices could also affect prices on substitutes for natural gas as a fuel or feedstock to fuels. Natural gas prices affect the production cost of fertilizer, which is a key input for producing cornstarch-based ethanol fuel. Also, a lower gas price would be likely to incentivize more R&D investment in bi-fuel vehicle technologies burning natural gas–hydrogen blends[32] although such bi-fuel vehicles are not commercially viable currently. It could also affect the commercial feasibility of electric vehicles and hydrogen fuel cell vehicles since natural gas could be used to generate electricity and is the most economic feedstock to produce hydrogen (Lee et al. 2012).

The Manufacturing/Industrial Sector

As shown in Fig. 3, the manufacturing sector is actually the largest user of natural gas. At the same time, US manufacturing has been declining in recent decades, as a result of increased international competition, recession, and a gradual shift toward the service industry. Such changes in economic activities, together with improvement in energy efficiency, have led to a reduction of 20 % in natural gas consumption from this sector during the past 15 years (Lipschultz 2012) and even more going back to the early 1970s. However, the reduction in natural gas prices brought about by shale development has stimulated a series of expansion announcements in manufacturing, especially in gas-intensive industries like petrochemicals and fertilizers. Some big European manufacturers are planning to move their production plants back to the USA to take advantage of cheaper natural gas. For example, Huntsman Corp., a chemical company that used to spend 90 % of its discretionary growth capital outside of the USA now is said to be spending 70 % within the UASA because of cheap gas (Johnson and Tullo 2013). And the boom in natural gas has also made the USA attractive to foreign investments, as evidenced by a recently announced plan by German-based BASF to build a chemical plant in Louisiana.[33]

The manufacturing sector could benefit from abundant natural gas in several additional ways. First, equipment manufacturers and construction material providers could experience a boost in demand due to shale gas expansion. Second, companies in the petrochemical industry would benefit from the cost reduction from cheaper raw materials and energy input, which will subsequently pass on at least some of the cost advantage to downstream sectors (e.g., plastic and rubber) through the value chain. Lastly, the growth in income, employment, and tax revenue

[32] Such as Hythane™(20 % H_2 by volume) or "HCNG" (30 % H_2 by volume) blends.

[33] BASF, "Governor Jindal and BASF Dedicate Methylamines Plant in Geismar", http://www.basf.com/group/corporate/en_GB/news-and-media-relations/news-releases/news-releases-usa/P-10-0109

Table 5 Natural gas consumption and direct output gain in manufacturing industries

Manufacturing industries	Annual natural gas consumption	Share of natural gas in total energy consumption	Direct industry output gain in 2015–2020 (%)	Direct industry output gain in 2015–2020 (2010$ billions)
Chemicals[a]	1.7 Tcf	33 %	14.5 %	70.2
Paper	460 Bcf	20 %	2.2 %	3.7
Plastic and rubber products	125 Bcf	38 %	17.9 %	33.28
Glass	150 Bcf	53 %	3.3 %	0.656
Iron and steel	375 Bcf	35 %	4.4 %	5.03
Aluminum	180 Bcf	49 %	7.6 %	1.69
Foundries	120 Bcf	44 %	2.4 %	0.617
Fabricated metal products	235 Bcf	61 %	1.8 %	5.81

Source: American Chemistry Council (2012)
[a]This excludes pharmaceuticals

resulting from a manufacturing renaissance could stimulate the broader economy by increasing demand in consumption and government spending.

The American Chemistry Council (2012) examined the potential economic benefits of shale gas development among eight energy-intensive manufacturing industries in the USA. The study, which was based on an economy-wide input–output model (the IMPLAN model), assessed the economic gains at three levels: direct effects, indirect effects, and induced effects.[34] According to the study, the direct effects include $121.0 billion in additional industry output and $72 billion in capital investment measured in 2010 constant US dollars over the period 2015–2020, which is equivalent to a 7.3 % gain above the reference output level. The indirect effect (to supplier industries) was estimated to result in an additional $143.8 billion growth in economic output. Such economic expansion will subsequently lead to increased demand in other sectors of $76.8 billion induced economic gain through income and tax growth. Table 5 provides estimates of direct industry output gains for each of the eight manufacturing industries along with their natural gas consumption. These gains varied from 1.8 % to 17.9 % based on each industry's baseline output. Among these eight industries, chemicals and plastic and rubber products are the most important contributors to these gains (about 85 % of total estimated output gain).

A number of other research institutes and consulting firms have come up with their own estimates of the economic and employment benefits of the shale gas boom on the US manufacturing industry. A report from Price Waterhouse Coopers ([PwC] 2011,

[34]Here, *direct effects* refer to the output and employment effects generated by the sector itself; *indirect effects* refer to such effects supported by the sector via purchases from its supply chain; and *induced effects* refer to the employment and output supported by the spending of those employed directly or indirectly by the sector (American Chemistry Council 2012).

Table 6 Announced manufacturing projects in the USA related to shale gas availability

Industry	Company	Project	Location	Announced investment	Time of announcement
Petrochemical[a]	Methanex Corp.	Methanol manufacturing plant moved from Chile	Ascension Parish, Louisiana	$550 million	Jan. 2012
Petrochemical[b]	Williams	Expansion of an ethylene plant	Geismar, Louisiana	$350–400 million	Sep. 2011
Petrochemical[c]	Dow Chemical	A new ethylene plant	Freeport, Texas	N/A	Apr., 2012
Textile[d]	Santana Textiles LLC	Denim plant	Edinburg, Texas	$180 million	Jul., 2008
Fertilizer[e]	CF Industries	Expansion of a nitrogen fertilizer manufacturing complex	Donaldsonville, Louisiana	$2.1 billion	Nov. 2012
Fertilizer[f]	Orascom Construction Industries	Nitrogen fertilizer production plant	Southeast Iowa	$1.4 billion	Sep. 2012

Source: Media coverage

[a]Office of the Governor, State of Louisiana. "Governor Jindal, Methanex Announce $550 Million Methanol Plant". Jul 25, 2012. http://gov.louisiana.gov/index.cfm?md=newsroom&tmp=detail&articleID=3545

[b]Williams, Inc. "Williams Expanding Geismar Facility to Serve Petrochemical Industry". Sep 20, 2011. http://www.energy.williams.com/profiles/investor/ResLibraryView.asp?ResLibraryID=47352&GoTopage=5&Category=1799&BzID=630&G=343

[c]The Dow Chemical Company. "Dow to Build New Ethylene Production Plant at Dow Texas Operations". April 19, 2012. http://www.dow.com/texas/freeport/news/2012/20120419a.htm

[d]Edinburg Politics. "Santana Textiles Corporation of Brazil to build $180 million manufacturing plant in Edinburg". July 4, 2008. http://www.edinburgpolitics.com/2008/07/04/santana-textiles-corporation-of-brazil-to-build-180-million-manufacturing-plant-in-edinburg/

[e]Louisiana Economic Development. "CF Industries Announces $2.1 Billion Expansion In Donaldsonville". Nov 1, 2012. http://www.louisianaeconomicdevelopment.com/index.cfm/newsroom/detail/217

[f]Wall Street Journal. "Egyptian Bets $1.4 Billion on Natural Gas—In Iowa". Sep 5, 2012. http://online.wsj.com/article/SB10000872396390443589304577633932086598096.html

p. 1) estimates that "lower feedstock and energy costs could help U.S. manufacturers reduce natural gas expenses by as much as $11.6 billion annually through 2025."

Table 6 shows a list of new or expanded projects announced in recent years that are partially credited to the shale gas boom. According to American Chemistry Council (2014), more than 148 chemical projects have been publicly announced as of February of 2014, totaling $ 100.2 billion in new investment. Texas, Louisiana, Oklahoma, and Pennsylvania—which are big shale gas-producing states—are on top of the list in terms of the project locations. These projects, once brought online, would add significant base-load demand for natural gas. For example, an ammonia plant with a production capacity of 1,500 t per day could consume 44 MMcf of natural gas per day, which is equivalent to the estimated gas consumption of 165,000 CNG vehicles (Lipschultz 2012).

However, unlike the electric power sector, where existing gas-fired plants are present for fuel switching, most of these projects require a large amount of investment with a payback period of up to 5 years before they can generate positive cash flows. Therefore, the economic benefits in manufacturing are more vulnerable to gas price volatility and to underlying concerns about LNG exports boosting price.

Currently, DOE has granted seven licenses for LNG to be exported to countries that lack a free-trade agreement with the USA.[35] More approvals are likely. While such exports may raise natural gas prices and reduce profits for gas feedstock users, the loss was estimated to be manageable (NERA Economic Consulting 2012). The costs from liquefaction and transportation to Japan and China could add up to about $5.50 per Tcf (Johnson and Tullo 2013). And landed LNG rates are typically linked to oil prices, with an oil price of $100/bbl translating into a landed LNG rate of $12.00–$15.50/mmBtu (Lipschultz 2012), which would probably keep the US gas price well below the prices in gas-importing countries.

The subsections below take a closer look at three industries where the impacts of cheap gas availability are believed to be the most prominent: petrochemical, fertilizer, and steel production.

Petrochemicals

The petrochemical industry is one of the largest natural gas consumers, where natural gas and natural gas liquids (NGLs) are used as fuels and as feedstock. It is estimated that chemicals and petroleum refining account for 46 % of total industrial gas consumption (Lipschultz 2012). Important NGLs for the petrochemical industry include ethane, propane, and butane. Ethane and propane are the primary feedstocks used in the USA to produce ethylene, which is a key component in plastics and one of the world's most common chemical building blocks. An expansion in the production capacity of ethylene will probably boost production from a wide variety of manufacturing industries, such as electronics, clothing, and packaging, therefore leading to far-reaching impacts on the entire manufacturing sector.

An American Chemistry Council (2011) study focusing on the petrochemical industry indicates plans for investment of $16.2 billion to build new petrochemical

[35] Only four free-trade agreement countries (South Korea, Australia, Mexico, and Canada) are big natural gas consumers; of these, only South Korea is a major LNG importer. The world's largest LNG importers, such as Japan, China, and the UK, are non-free-trade-agreement countries (Johnson et al. 2013).

and derivatives capacity arising from the availability of cheap gas over several years; this is projected to increase US ethane capacity by about 25 %. Similarly, PwC (2012, p. 3) estimates that NGL production is "expected to increase more than 40 % over the next 5 years, reaching more than 3.1 MMBD [million barrels per day] in 2016," and the US chemical industry investment of $15 billion will have increased its ethylene production capacity by 33 %.

A number of petrochemical manufacturers, including Dow Chemical, Formosa Plastics, Chevron Phillips Chemical, and Bayer Corp., have announced plans to build new plants or expand existing capacity for ethane and ethylene production partly because of the availability of shale gas feedstock (PwC 2011). Dow Chemical (2011, p. 11) stated that its "investments on the U.S. Gulf Coast will increase [its] U.S. ethylene production capabilities by 20 percent over the next three years." An analysis based on economic cost models of petrochemical products shows that, as the price of natural gas falls from $12.5/mmBtu to $3.00/mmBtu, the estimated price of ethylene declines from $1,009/t to $323/t, with the prices of polyethylene and ethylene glycol falling by a similar degree (PwC 2012). Such a cost reduction will give the US chemical manufacturers a significant cost advantage over its international competitors.

In addition, petrochemical manufacturers are shifting from petroleum-based feedstocks to their gas-based substitutes. The petrochemical producers who were using naphtha (a generic term for a variety of petroleum refining products) as a feedstock for ethylene production, primarily located in Europe and Asia, are shifting to ethane to reduce costs, which has caused the naphtha prices to fall in the USA and international markets (Pirog and Ratner 2012). It is expected that research and development efforts leveraging ethylene-based chemistries that replace petroleum-based products will surge in the coming years. Finally, a variety of downstream manufacturing sectors will subsequently benefit from the availability of cheaper chemical raw materials, which are used in an estimated 90 % of all manufactured products and may replace higher-cost materials such as metals, glass, and leather (PwC 2012).

Fertilizers

Natural gas is used to produce ammonia, which serves as the primary ingredient in most nitrogen fertilizers and is an essential ingredient in many finished phosphate fertilizers; the cost of ammonia makes up about 70–90 % of the estimated production cost faced by nitrogen-based fertilizer producers (Pirog and Ratner 2012). Although the USA is the fourth-largest producer of ammonia in the world, US ammonia production capacity shrank by 40 % during the 2000s, before the shale gas boom changed the situation.[36] Many ammonia plants were moved overseas or closed because of the increasing natural gas price for industrial users. However, the emergence of large, low-cost

[36]The Fertilizer Institute, "Natural Gas Access/ Supply", http://www.tfi.org/issues/energy/natural-gas-accesssupply

shale gas resources has reversed the trend and brought significant cost savings to the industry. With the high demand for fertilizers in the past few years, the cost savings brought by cheap natural gas were mainly reaped by the producers rather than the consumers (Pirog and Ratner 2012). CF Industries, a major manufacturer and distributor of fertilizer products in the USA, reported a more than threefold increase in gross margin of its nitrogen segment from 2009 ($784 million) to 2011 ($2,563 million; CF Industries 2012). Accordingly, new investment decisions to expand capacity by both domestic and foreign producers have been announced recently. Nevertheless, this industry is cautious because of losses incurred in investments made when gas prices were low in the past and then rose dramatically (Pirog and Ratner 2012).

Steel Production

The benefit of expanded shale gas development for steel production comes from two factors: the increase in product demand caused by higher demand for drilling equipment and the decrease in operating costs due to cheaper natural gas. For example, US Steel (the largest US steelmaker) saw a 17 % increase in production of tubular goods used in oil and gas drilling and transmission facilities in 2011 (Miller 2012). The increasing demand for drilling equipment, together with the demand-driven increase in steel prices and downturn in costs, has increased the company's profits.

In addition, the steel industry can benefit directly from fuel switching, from coal to natural gas, both by directly replacing coal with natural gas in manufacturing processes and by experiencing lower electricity prices made possible partly by the increased gas supply. For example, Nucor and US Steel have both indicated interest in investing in the use of natural gas for direct reduced iron production (PwC 2011)—a process that has traditionally used coal to create the requisite "reducing gas."

However, the cost reduction from switching to natural gas from coal falls into the range of $8–$10/t, compared to the overall steel production cost of around $600/t (Miller 2012; Pirog and Ratner 2012). Therefore, the cost reduction effect is less important to the industry than the upward shift in demand.

Looking into the future, as projected by EIA (2014), natural gas use in the industrial sector increases by 22 % from 8.7 quadrillion Btu per year in 2012 to 10.6 quadrillion Btu per year in 2025. Industrial production from the bulk chemical industries is projected to grow 3.4 % per year from 2011 to 2025, partly powered by the increased production of natural gas and NGLs (EIA 2014).

Conclusion

The natural gas revolution is already having profound effects on the electric power and manufacturing sectors of the US economy and is likely to have growing effects on the transportation sector in the future, particularly the heavy-duty truck market, which is already seeing high market share of LNG to fuel trash trucks.

The power sector is seeing an increasing amount of natural gas used to generate electricity, as a result of using previously built, but unused capacity. Further growth in natural gas use is likely but at a slower rate. Partly this is due to using up excess generation, and partly this is a result of renewable mandates at the state level.

The manufacturing sectors that are dependent on natural gas feedstock or can substitute gas for other fuels have definitely been boosted by low natural gas prices, and there is no reason that this trend will not continue for the foreseeable future both through increased foreign and domestic investment.

As for transportation, we find that penetration of natural gas in the heavy-duty truck market makes economic sense, but that without breakthroughs in using natural gas as a feedstock for liquid fuels, there is unlikely to be major penetration in the light-duty vehicle market.

Not discussed here is the longer run outlook where the burning of fossil fuels becomes potentially incompatible with meeting CO_2 emission reduction goals. One can imagine that natural gas can substitute for coal in generation or manufacturing as a bridge to a low carbon future if fugitive methane emissions can be more tightly controlled. However, the substitution of gas for oil is more problematic from a CO_2 emission perspective since lifecycle CO_2 emissions from oil are significantly below those of coal and, at least in trucking, liquefaction activities increase the carbon footprint of natural gas. If demand for natural gas from the power sector falls, these other sectors may benefit even more from cheap natural gas.

Also not discussed above is the residential and commercial use of natural gas. Demand for this sector is expected by EIA (2014) to be essentially flat. But IHS (2014) finds that there are opportunities to expand local distribution networks, such as in the Northeast USA, to reach more consumers. However, outdated rules governing the approval of investments in this regulated sector—basically limiting investments to those with a short-payoff period on the assumption that natural gas prices are more volatile than they are now expected to be—are hindering planning for such projects.

References

Alternative Fuels Data Center (AFDC) (2011) Alternative fueling station total counts by state and fuel type. Alternative Fuels & Advanced Vehicles Data Center in the U.S. DOE. Retrieved 20 Apr 2011. Available at http://www.afdc.energy.gov/afdc/fuels/stations_counts.html

American Chemistry Council (2011) Shale gas and new petrochemicals investment: benefits for the economy, jobs, and U.S. manufacturing. American Chemistry Council, Economics and Statistics Department, Washington, DC

American Chemistry Council (2012) Shale gas, competitiveness and new U.S. investment: a case study of eight manufacturing industries. American Chemistry Council, Economics and Statistics Department, Washington, DC

American Chemistry Council (2014) U.S. chemical investment linked to shale gas reaches $100 Billion. American Chemistry Council, Economics and Statistics Department, Washington, DC

Bryce R (2011) Ten reasons why natural gas will fuel the future. Energy policy and the environment report no. 8, Apr. Manhattan Institute for Policy Research, New York. Available at http://www.manhattan-institute.org/html/eper_08.htm

Burtraw D, Palmer K, Paul A, Woerman M (2012) Secular trends, environmental regulations, and electricity markets. Discussion paper 12–15. Resources for the Future, Washington, DC

Celebl M, Graves F, Russell C (2012) Potential coal plant retirements: 2012 update. Discussion paper. Oct. The Brattle Group, Washington, DC

Center for Climate and Energy Solutions (C2ES) (2012) Natural gas use in the transportation sector. Center for Climate and Energy Solutions, Arlington, Available at http://www.c2es.org/publications/natural-gas-use-transportation-sector

CF Industries (2012) CF industries 2011 annual report: our global table. CF Industries Holdings, Deerfield

Chemical D (2011) 2011 annual report: welcome to solutionism™. The Dow Chemical Company, Midland

Cianca D (2011) Natural gas passenger vehicles coming to grand junction: Honda's civic GX expected locally by early fall. KJCT8.com of Pikes Peak Television Inc., 7 Apr. Available at http://www.kjct8.com/news/27473206/detail.html

Energy Information Administration (EIA) (2008) Natural gas weekly update. Department of Energy, Washington, DC

Energy Information Administration (EIA) (2012a). Annual energy review 2011. DOE/EIA-0384(2011). U.S. Department of Energy, Washington, DC

Energy Information Administration (EIA) (2012b) Electricity generation from coal and natural gas both increased with summer heat. Available at http://www.eia.gov/todayinenergy/detail.cfm?id=8450#

Energy Information Administration (EIA) (2012c) Annual energy outlook 2012. DOE/EIA-0383(2012). U.S. Department of Energy, Washington, DC

Energy Information Administration (EIA) (2014) Annual energy outlook 2014 early release overview. U.S. Department of Energy, Washington, DC

Energy Information Administration (EIA) (n.d.) Primary energy consumption by source and sector, 2012. U.S. Department of Energy, Washington, DC. http://www.eia.gov/totalenergy/data/monthly/pdf/flow/css_2012_energy.pdf

Energy Vision (2013) Tomorrow's trucks: leaving the era of oil behind. Energy Vision, New York

Federal Energy Regulatory Commission (2013) FERC authorizes information sharing by gas pipelines, electric utilities. News release, docket no. RM13-17-000. Federal Energy Regulatory Commission, Washington, DC. 15 Nov 2013. Available at https://www.ferc.gov/media/news-releases/2013/2013-4/11-15-13.asp#.U1NgVV5j5g0

Fraas AG, Harrington W, Morgenstern RD (2013) Cheaper fuels for the light duty fleet: opportunities and barriers, Forthcoming discussion paper. Resources for the Future, Washington, DC

Gladstein, Neandross & Associates (2011) NGV roadmap for Pennsylvania jobs, energy security and clean air. Marcellus Shale Coalition, Canonsburg, Available at http://marcelluscoalition.org/wp-content/uploads/2011/04/MSC_NGV_Report_FINAL.pdf

Goulding AJ, Zhang Y, Hoang V (2011). Seeing past the hype: why natural gas vehicles (NGVs) may be more cost effective than electric vehicles (EVs) for fighting climate change. 18 Feb. London Economics International LLC, Boston. Available at http://www.londoneconomics.com/pdfs/LEI_NGV_paper_updated_Final.pdf

Gray R (2011). 2011 NatGas act introduced in congress. School transportation news, 6 Apr. Available at http://www.stnonline.com/home/latest-news/3257-2011-natgas-act-to-be-introduced-to-congress

Hesterberg T, Bunn W, Lapin C (2009) An evaluation of criteria for selecting vehicles fueled with diesel or compressed natural gas. Sustain Sci Pract Policy 5(1):20–30

IHS CERA (2014) Fueling the future with natural gas: bringing it home. Available at http://www.fuelingthefuture.org/assets/content/AGF-Fueling-the-Future-Study.pdf

IHS Global Insight (2010) Presentation by Mary Novak on U.S. Energy Outlook at annual Energy Information Administration conference, 6–7 Apr, Washington, DC

Johnson J, Tullo AH (2013) Chemical and gas suppliers battle over LNG exports. Chem Eng News 91(10):9–13

Kell-Holland C (2009) Payoff—one company rolled the dice on LNG and reaped the reward. LandLine Magazine, Mar/Apr: 53. Available at http://www.cleanenergyfuels.com/pdf/LNG_landline_apr09.pdf

Knittel CR (2012) Leveling the playing field for natural gas in transportation, Discussion paper 2012-03. Brookings Institution, Washington, DC

Lee A, Zinaman O, Logan J (2012) Opportunities for synergy between natural gas and renewable energy in the electric power and transportation sectors. Technical report NREL/TP-6A50-56324, Dec. National Renewable Energy Laboratory, Joint Institute for Strategic Energy Analysis, Golden

Lipschultz MC (2012). Historic opportunities from the shale gas revolution. KKR report, Nov. Kohlberg Kravis Roberts & Co. L.P., New York

Logan J, Heath G, Macknick J, Paranhos E, Boyd W, Carlson K (2012) Natural gas and the transformation of the U.S. energy sector: electricity. Technical report, NREL/TP-6A50-55538. The Joint Institute for Strategic Energy Analysis, Denver

Macmillan S, Antonyuk A, Schwind H (2013) Gas to coal competition in the U.S. power sector. OECD/IEA, Paris. Available at http://www.iea.org/publications/insights/CoalvsGas_FINAL_WEB.pdf

Massachusetts Institute of Technology (2009) The future of natural gas, an interdisciplinary study of MIT. MIT Energy Initiative, Cambridge, MA

Miller JW (2012) Steel finds sweet spot in the shale. Wall Str J, 26 Mar. Available at http://online.wsj.com/article/SB10001424052702304177104577305611784871178.html

Mufson S (2012) The demise of coal-fired power plants. The Washington Post, 23 Nov. Available at http://articles.washingtonpost.com/2012-11-23/business/35508510_1_coal-plant-coal-prices-coal-mines

NERA Economic Consulting (2012) Macroeconomic impacts of LNG exports from the United States. NERA Economic Consulting, Washington, DC

Pirog R, Ratner M (2012) Natural gas in the U.S. economy: opportunities for growth. Congressional Research Service, Washington, DC

Port of Los Angeles and Port of Long Beach (2011) San Pedro Bay Ports Clean Air Action Plan. Clean Air Action Plan (CAPP) Implementation progress report, quarterly report—1st quarter 2011. Available at http://www.cleanairactionplan.org/civica/filebank/blobdload.asp?BlobID=2518

PricewaterhouseCoopers (PwC) (2011) Shale gas: a renaissance in U.S. manufacturing? Dec. PwC, New York

PricewaterhouseCoopers (PwC) (2012). Shale gas: reshaping the U.S. chemicals industry. Oct. PwC, New York

Taschler J, Content T (2011) Interest soars in natural gas vehicles: as oil prices rise, more business fleets could use alternative fuel. The Milwaukee J Sentin, 9 Apr. Available at http://www.jsonline.com/business/119517879.html

Taylor GD, DuCote WG, Whicker GL (2006) Regional fleet design in truckload trucking. Transp Res Part E 42(3):167–190

Turner S (2010) Natural gas vehicles: what's here and what's coming. Presented at the waste-to-wheels: building for success conference. 1 Dec, Columbus. Available at http://www1.eere.energy.gov/cleancities/pdfs/ngv_wkshp_turner.pdf

U.S. Environmental Protection Agency (2012) Inventory of U.S. greenhouse gas emissions and sinks 1990–2010. EPA 430-R-12-001. Apr. U.S. Environmental Protection Agency, Washington, DC

U.S. Environmental Protection Agency and the U.S. Department of Transportation's National Highway Traffic Safety Administration (2011) Greenhouse gas emissions standards and fuel efficiency standards for medium- and heavy-duty engines and vehicles. Federal Register 76(179):57106–57513

Weiss J, Bishop H, Fox-Penner P, Shavel I (2013) Partnering natural gas and renewables in ERCOT. The Brattle Group, Cambridge, MA

White House (2012) Fact sheet: all-of-the-above approach to American Energy. Office of the Press Secretary, 7 Mar. Available at http://www.whitehouse.gov/the-press-office/2012/03/07/fact-sheet-all-above-approach-american-energy

White House (2013) Fact sheet: President Obama's blueprint for a clean and secure energy future. Office of the Press Secretary, 15 Mar. Available at http://www.whitehouse.gov/the-press-office/2013/03/15/fact-sheet-president-obama-s-blueprint-clean-and-secure-energy-future

Achieving Balance Between Economic, Sociodemographic, Environmental, and Regulatory Concerns: A Shale Gas Perspective

Clifford A. Lipscomb, Sarah J. Kilpatrick, Yongsheng Wang, and William E. Hefley

Abstract As with the development of any new industry, shale gas development poses opportunities and challenges. Economically, the benefits related to jobs and tax revenues must be balanced against potential costs including environment, property value, litigation, and financing. The shale energy industry must proactively face these challenges and downstream effects. We summarize these challenges and downstream effects and discuss the much needed future research that is related to these difficult questions and could provide comprehensive understanding.

Introduction

Previous chapters have shown that shale gas development can be considered from multiple perspectives. Some of these perspectives are economic in nature, including increased output and employment opportunities in the gas sector. Economic impacts are disaggregated into direct, indirect (i.e., local industries buying goods and

C.A. Lipscomb (✉)
Greenfield Advisors LLC, 106 N. Bartow Street, Cartersville, GA 30120, USA
e-mail: cliff@greenfieldadvisors.com

S.J. Kilpatrick
Greenfield Advisors LLC, 13517 35th Avenue S, Tukwila, WA 99168, USA
e-mail: sarah@greenfieldadvisors.com

Y. Wang
Washington and Jefferson College, 60 South Lincoln Street, Washington, PA 15301, USA
e-mail: ywang@washjeff.edu

W.E. Hefley
Joseph M. Katz Graduate School of Business and College of Business Administration, University of Pittsburgh, Mervis Hall, Pittsburgh, PA 15260, USA
e-mail: wehefley@katz.pitt.edu

W.E. Hefley, Y. Wang (eds.), *Economics of Unconventional Shale Gas Development*, Natural Resource Management and Policy 45,
DOI 10.1007/978-3-319-11499-6_10

services from other local industries), and induced (re-spending of income) multiplier effects that can filter through the local and regional economies. Other perspectives are noneconomic (e.g., the sociodemographic, environmental impact and regulation of shale gas development). Regardless of the perspective, there are trade-offs that must be considered when a policymaker or regulatory authority makes a decision regarding some aspect of shale gas development, whether that decision is to fund local infrastructure improvements or to ban shale gas development altogether (as the State of Vermont has done and other communities are considering as of this writing).

Shale gas development poses some unique issues. How the industry resolves these issues is rooted in the degree of the perception that there *are* issues to address (e.g., climate change, safety and health concerns, and property value impacts due to shale gas development) and that these issues may be counteracted by increased output and employment opportunities, increased tax revenues at the municipal level, and additional available gas resources to sell on the open market. Often the costs of shale gas development affect some persons and entities, while the benefits are received by different persons and entities. What decisions need to be made to find the balance between the costs imposed on some groups and the benefits that accrue to others? Should rules and regulations be put in place to minimize the costs on state and/or local governments? What about the costs imposed on citizens and homeowners? Then, more broadly, how does shale gas development affect the US energy portfolio and security? What are the impacts on the world?

Challenges in the Natural Gas Industry

The questions posed above motivate a discussion of the challenges that face the natural gas industry broadly. Unconventional shale gas development through hydraulic fracturing ("fracking") is not the first boom in the oil/gas industry. Many people recall the pump jacks and derricks of yesteryear that once were everywhere there was oil. Some of these pump jacks and derricks are still around today and functional. Others have been abandoned as the industry has progressed technologically. Of the many things to consider when we speak to the challenges of the oil/gas industry, we should think of where it has been and where it is headed in order to put things into perspective.

In reviewing both the oil/gas booms of the past and the shale oil and gas boom of recent years, similar things occurred: workers migrated into new areas where they took up either temporary or permanent residency; housing was often in short supply when they arrived; drilling brought more oil/gas-related jobs; and those jobs spurned indirect and induced jobs. In other words, we are not really dealing with anything new, yet there is quite a bit of controversy reported on the topic.

Perhaps the only challenges we think about today that are different from the oil/gas pioneering days are public policy and the environment. In fact, the US Environmental Protection Agency did not begin its operations until 1970, a fair amount of time after most of the initial oil/gas booms. While public policy and the environment may have been less of a prevailing concern in years gone by, these are probably the biggest factors in today's industry. In the sections below, we discuss the impacts on the oil/gas industry while recapping the findings highlighted in previous chapters and expand these where able with additional and suggested research.

Economic Impacts

The fact that there are economic incentives related to unconventional shale gas development, and fairly significant ones at that, is the reason why there is a struggle between various levels of government and the real or even perceived environmental trade-offs. The previous chapters in this book have shown that employment, income, and housing impacts are probably the most studied of these impacts, followed by a lesser extent the sociodemographic effects.

Job creation is probably the most visible and likely the most economically beneficial gain from shale gas development. In 2009, the economic impact to just the Marcellus Shale region in Pennsylvania was between 23,385 and 23,884 jobs, making up about $1.2 billion dollars in labor income and with an additional value add to the Pennsylvania economy of $1.9 billion (Kelsey et al. 2011). Though the various studies calculate their numbers differently, that same year, West Virginia employed 9,869 persons in the oil/gas industry paying $551.9 million in wages with a total value impact to the state of over $12 billion creating approximately 24,400 jobs (7,600 of which were generated that year) (Higginbotham et al. 2010). In 2012, based on a group of 14 Texas counties producing oil and gas, the total economic impact was $46 billion, which supported 86,000 jobs. When combined with an additional six nearby nonproducing counties, the Eagle Ford area had an impact of $61 billion and supported 116,000 jobs (Oyakawa et al. 2013). Jobs are expected to increase in all of the areas noted above based on current projections.

It is important to understand the trade-offs of *not* fracking. While a significant amount of research has looked at the economic impacts of fracking, a recent study conducted by researchers at the University of Colorado Boulder (Wobbekind and Lewandowski 2014) highlights the economic disincentives to a fracking ban in the State of Colorado. The results of this research indicate that if a fracking ban were to take place in 2015 with the activities in place allowed to deplete, then in the long term (2015–2040), there would be 93,000 fewer jobs and $12 billion less in gross domestic product (GDP). In the first 5 years of a ban, 68,000 of those jobs would not be created, which corresponds to $8 billion in GDP.

While jobs have a large economic impact from shale gas development, we must not forget about lease payments and royalties which incentivize owners to allow the development but also provide these owners with income. Probably one of the more interesting findings in recent studies is the savings rates from both leasing and royalty income in northeastern Pennsylvania. Studies (Kelsey et al. 2012a; b; c; d; e) from the Marcellus Shale Education and Training Center (a collaboration of

Pennsylvania College of Technology and Penn State Extension) indicate that in the Pennsylvania counties of Bradford, Sullivan, Susquehanna, Tioga, and Wyoming, the savings rate on leasing dollars is approximately 55 % (marginal savings) with income from royalties being saved a rate of approximately 66 % (marginal savings). With the 2012 average savings rate as a percentage of disposable personal income at 8.2 % (Bureau of Economic Analysis 2014), this is well above the savings rate that might be expected. Typical taxation on leasing income is around 17.52 %. Of the leasing income actually spent, approximately 9.02 % is spent on motor vehicles, 5.15 % on real estate, 4.36 % on farming, 1.75 % on real estate improvement, 1.65 % on healthcare and insurance, and the rest is spent on vacations, travel, entertainment (0.36 %), consumer goods (0.2 %), food (0.01 %), and with a remaining 4.79 % spent on other items not specifically categorized. Understanding how these monies from fracking activities are spent is key to understanding how local markets will respond to fracking activities.

Often no consideration is given to the trade-off of one type of land use over another or the conditions under which the highest and best use of land changes. Farmers can continue to plant crops on the land with shale gas wells, but infrastructure such as access road and drilling pads could separate large pieces of an estate and make farm planning more challenging. While housing is often in short supply as new jobs are created, studies show that workers often only temporarily migrate or even commute to new areas. This may be due to the relatively short time it takes to get a well to the production phase. Research from the University of Pittsburgh (Hefley et al. 2011) indicates that site construction typically takes 5 days, drilling 18–21 days, fracking 1–3 days, and well completion from 10 to 15 days, at which point a well is producing until it stops or production is halted. This is a short time period for workers to relocate for just one well, albeit if an abundance of this type of work is available in an area, relocation would be more likely, and even more feasible if the housing supply is available.

In Williston, North Dakota, with a population of 18,500, the creation of 75,000 new jobs has the job and housing markets in disproportionate supply. News articles (Gelber 2014; Wood 2013) indicate that hotels are renting for "affordable" rates of $700 a night and that while oil/gas workers may be making six figures, many are living in the local Walmart parking lot. Rents are noted to be four times higher than in the 5 years prior, and houses that were selling for $60,000 are noted as being listed for $200,000. Although this may be an uncommon example, it reminds us of the economic considerations that municipalities face in an economic boom.

If municipalities wish to attract these workers permanently, they will need to ensure their planning provides adequate housing/building permits; otherwise these workers are likely to stay in a hotel or other temporary housing when they are working nearby or commute from areas with available housing. Research indicates that oil/gas workers often come from other states (e.g., workers from Texas in Pennsylvania). This is indicative of a lack of trained/available workers in the immediate market. If regional markets want to ensure that they receive the tax revenues from income, it is important to develop training and educational programs for these types of jobs, or else these monies will leak to other areas. Adequate housing supply, be it temporary or permanent, is important to ensure these jobs stay local and that the growth from these economic "booms" does not leave when production ends.[1]

There is not a lot of research available on how property values change simply due to the presence of fracking activities (i.e., where contamination does not exist or is not known to exist). Boxall et al. (2005) found a negative impact on residential property value from proximity to traditional oil/gas facilities (this article was written before fracking became so prevalent). Gopalakrishnan and Klaiber (2012) studied the effects of shale gas exploration on property values in Washington County, Pennsylvania. Their results indicated that properties are adversely affected by proximate gas exploration sites/wells and that this effect dissipates with time and distance (the effect seems to disappear at 2 miles from a gas exploration site). They also found that adverse impacts are larger for properties with well water as their main drinking water source and are larger for properties near agricultural land. Similarly, Muehlenbachs et al. (2014) utilized a hedonic model to determine if proximity to both vertical and horizontal gas wells creates a difference in property values through water supply. Their results indicate that property values increased for houses with "piped water" (i.e., public water) due to the positive economic impacts of natural gas wells. However, for those houses with "groundwater" (i.e., private wells), the presence of gas wells created a net decrease in housing values. These differences may be attributable to the perceived risk of potential groundwater contamination, which is largely related to the common nature of drilling and not necessarily fracking.

A related housing issue is whether the presence of unconventional shale gas development sites affects the ability of homeowners to obtain a mortgage and/or homeowners' insurance on a property. These concerns stem from the potential environmental hazards and risks that may occur from shale gas activities. If environmental contamination or risk of contamination is present (e.g., if drilling activities are present on a property), some lenders will not originate mortgage loans on those residential properties; and insurance companies may not cover an incident that damages or contaminates a property that has these activities (Radow 2011; Nationwide Mutual Insurance Company 2012). While the lack of mortgageability and insurability may seem a minor consequence, the lack of mortgageability means a potential buyer would need to pay cash or other equivalents for the property, or the owner would not be able to sell the property because a mortgage would not be allowed. In addition, the inability to insure a property prevents a homeowner or investor from protecting their investment. If the owner is holding this real estate for investment purposes or as a type of generational wealth, then these concerns may lock the owner into the property without an exit strategy. Of course, if homeowners happen to be the owners of underground resources, they will be compensated with royalties. However, there is an issue of a split estate when mineral owners are not the property

[1] A related concern is the impact of the shale gas (and oil) boom that unconventional development has had on local airports. The recent increases in commercial aviation activities in Minot, Williston, and Dickinson (North Dakota) have led to tapped out facilities and the need for larger terminals (O'Donnell 2014).

owners. In this situation, the compensation of access fee for surface property owners is minimal compared to royalties.

In addition to the economic variables that determine whether a shale gas well is developed and the impacts that occur when drilling and fracking exist, there are noneconomic impacts that occur as a direct or indirect result of shale gas development. These impacts can be grouped into sociodemographic, environmental, and regulatory impacts.

Sociodemographic Impacts

While most would not think of the social impacts that shale gas development can have, it is an important component to consider. With a large amount of shale gas exploration and drilling being done in rural areas that were previously quiet and often agricultural in nature, the nearby switch to industrial activities such as these can have an overwhelming effect. For residents whose lands were once farmed, there may be additional trade-offs for the use of the surface rights and for generational farmers that can be a life-changing event. Studies indicate that dairy farms in the Marcellus Shale may decline with drilling intensity (Adams and Kelsey 2012; Finkel et al. 2013), while other research (Frey 2012) shows that dairy farming in the Marcellus Shale could modernize, decline, or change to other agricultural forms with the influx of natural gas drilling monies. Regardless of what actually happens, change is expected and with change comes adjustment.

The addition of nonpermanent workers may also have effects in communities that are new to natural gas drilling activities. While additional people may be in the area, they may not have a vested interest in the local community. These workers often do not pay property or income taxes to the community in which they work, but still utilize or strain the utilization of local support services (such as roads and police forces). The only direct contribution they make may be in the form of sales taxes on temporary lodging (some communities call this a "hotel/motel tax") and any goods and services they may purchase while in the area. For small towns, this influx of "strangers" may cause concern or may alter the local culture. While these changes could be positive or negative, small communities in particular may struggle with these sociodemographic changes.

While we spoke to the impacts on housing previously, it is important to note that the lack of housing supply and the increase in rents or housing costs (temporary or permanent) may actually cause locals to leave the area due to lack of affordable housing; this would be especially true in areas with large rental markets. The ability to sell in a booming market may additionally cause those with investments to sell while the market is up. For instance, seniors with larger homes than they prefer may sell their homes and relocate, utilizing the profits from that sale as retirement income. Conversely, if the supply of housing were to grow and then oil/gas activities dissipate, the reverse may occur and owners might see their values collapse as residents (temporary or permanent) vacate the area. Ultimately, the sociodemographic composition of these communities may change, and communities may struggle with these transitions.

Additional sociodemographic concerns include access to education. Increased enrollment would typically be expected in K–12 education in areas where jobs are growing. However, as noted in Chap. 8, educators in Pennsylvania did not see a significant increase in enrollment for K–12 education. This is likely due to either lack of supply of housing (i.e., families cannot find a place to live) or due to the temporary nature of oil/gas work in a specific location for a duration of time.

UTSA researchers on Eagle Ford and the State of Texas show that secondary educational institutions are creating new programs specifically to train and educate the workforce for the oil/gas industry, according to Chap. 7 by Tunstall. The impact of shale gas development in the Eagle Ford Shale on the regional workforce's educational composition will continue to evolve as these programs come online at both community- and state-funded colleges and universities.

Environmental Impacts

Environmental concerns, or perhaps the lack of perceived environmental concern, are arguably the most controversial component in the shale gas industry. The scale of whether or not to conduct oil/gas production is often tipped by environmental concerns and counterbalanced with increased jobs and revenues.

A recent study by Muehlenbachs et al. (2014) demonstrates "that groundwater-dependent homes are negatively affected by shale gas development" and that the magnitude of these negative impacts is large in the 0–1.5 km range. In this study, for homes within 1 km of a well, the addition of another well pad at the margin has a statistically significant negative impact of 16.7 % on property values. This is an important finding because it takes into account the negative impact due to risk perceptions of groundwater contamination and the positive impact of lease payments and other adjacency impacts; these results suggest that the mere perception of groundwater contamination risk for homes proximate to shale gas wells can negatively impact property values. With more and more wells drilled around the country, it shows the urgency and importance of researching and educating the public about shale development activities.

A study by the National Academy of Sciences (National Academy of Sciences 2012) notes that energy development-related "induced seismic events" have been felt in Alabama, Arkansas, California, Colorado, Illinois, Louisiana, Mississippi, Nebraska, Nevada, New Mexico, Ohio, Oklahoma, and Texas. Results from their study indicate that hydrologic fracturing as presently implemented does not necessarily pose a high risk of induced seismic activity; however, injection for disposal of wastewater does pose some risk. In general their research indicates that the volume, rate, and temperature of injected/removed fluids as well as the pore pressure, permeability of surrounding layers, fault properties, the stress of the crustal conditions, distance from injection point, and the duration of the injection all play a vital role in understanding the associated risk of "induced seismic events."

One of the reasons that natural gas has boomed in recent years (outside of the technological advances in the sector) is that natural gas is considered to be a cleaner source of energy than coal.[2] According to the US Energy Information Administration, natural gas utilized as energy produces slightly more than half of the carbon dioxide that coal does.[3] In 2012, for instance, carbon emissions decreased by 3.8 %, in part due to a mild winter, advances in automotive energy efficiency, and the ongoing transition from coal to natural gas.[4] While natural gas may produce less carbon, the primary concern with greenhouse gases as it relates to natural gas development is methane. Research indicates that methane leakage may be the differential feature for the energy efficiency of natural gas development.[5] If the natural gas sector can keep methane leakage low, then natural gas will be more efficient than coal as a source of energy. However, if methane leakage is not controlled, we may be no better off as far as greenhouse gases are concerned.

At present, natural gas appears to be a viable and cleaner energy alternative to coal as long as the industry is regulated to limit methane leakage and potential impacts to drinking water. The proposed Clean Power Plan, if approved, will likely play an important role in this regulation.[6] In the future, we may find that natural gas was merely a stopgap between coal and even cleaner types of energy such as wind and solar power. New types of energy will likely come with their own trade-offs. Much like coal, natural gas is in finite supply and other types of renewable energy need to be explored even if natural gas becomes the primary source of clean energy for the global economy.

Regulatory Impacts and Policy

Taxation

Taxes are a huge component of the oil/gas industry. Where two markets may have similar production capabilities, the market with the more favorable taxation is likely to be chosen. With increased tax revenue, we believe it is important that

[2] "Natural Gas Really is Better than Coal." http://www.smithsonianmag.com/science-nature/natural-gas-really-better-coal-180949739/ (Retrieved July 27, 2014)

[3] US Energy Information Administration. "Frequently Asked Questions." How much carbon dioxide is produced per kilowatt hour when generating electricity with fossil fuels? http://www.eia.gov/tools/faqs/faq.cfm?id=74&t=11 (Retrieved July 27, 2014)

[4] US Carbon Dioxide Emissions Drop 3.8 Percent" http://www.livescience.com/40600-us-carbon-dioxide-emissions-drop.html (Retrieved July 27, 2014)

[5] Brandt AR et al. (2014) "Methane Leaks from North American Natural Gas Systems." Policy Forum. http://www.novim.org/images/pdf/ScienceMethane.02.14.14.pdf (Retrieved July 27, 2014)

[6] United States Environmental Protection Agency. "Clean Power Plan Proposed Rule." http://www2.epa.gov/carbon-pollution-standards/clean-power-plan-proposed-rule (Retrieved July 27, 2014)

policymakers use these tax dollars to protect local communities against any environmental risks that could arise and to maintain sustainable development. This would help ensure that the growth seen in these markets is sustained years from now as technology changes and shale gas production slows or moves to more productive or economically viable areas.

This raises an interesting question: Are communities using the additional tax dollars generated from the increase in shale gas development to prepare their communities for life after shale gas development? The one that comes to mind is Alaska and the Alaska Permanent Fund. When Alaskans realized that the robust income they were receiving from the Alaska Pipeline would not last forever, Alaska citizens "voted in 1976 to amend the constitution to put at least 25 % of the oil money into a dedicated fund…[which] would save money for future generations, which would no longer have oil as a source of income." Arguably the fund is successful because it was created by constitutional amendment, is very public, is not used as a bank, has performance oversight of managers, and receives legislative oversight.[7] In states where shale gas development is generating lots of additional tax dollars, policymakers will want to look to the Alaska Permanent Fund as an example.

Setback and Zoning Requirements

Most municipalities have zoning for rural, residential, commercial, and industrial areas or a combination of these. Zoning designations typically keep industrial activities away from residential areas, and in doing so actually help protect local residential property values. With fracking activities the same type of zoning or setback requirement should apply. Whether or not any type of environmental situation ever occurs, the truck traffic and noise from drilling makes this type of activity industrial in nature. Why then would a municipality allow fracking activities at any closer of a proximity than it would other industrial practices? Setback requirements and zoning play a vital role in the protection of the use and enjoyment of one's property. For example, Pennsylvania Supreme Court struck down portion of Act 13 and brought zoning jurisdiction of oil and gas activities back to local communities.[8]

[7] Alaska Permanent Fund Corporation. "Frequently Asked Questions." http://www.apfc.org/home/Content/aboutFund/fundFAQ.cfm (Retrieved July 25, 2014)

[8] "Pa. Supreme Court will not reconsider Act 13 decision." http://stateimpact.npr.org/pennsylvania/2014/02/21/pa-supreme-court-will-not-reconsider-act-13-decision/ (Retrieved on 07/24/2014)

Funding

A related topic is the source of capital to finance shale gas development. Most companies are using their own coffers to finance drilling rigs, lease agreements, etc. prior to the recovery of any oil or gas in a particular region. However, some oil/gas companies are raising funds through the sale of bonds. There is a relatively robust high-yield debt market in the USA. As quoted in Bloomberg Businessweek, a recent bond sale allowed Rice Energy, a natural gas producer with a low credit rating that plans to invest $1.2 billion in shale gas development in Pennsylvania's Marcellus Shale and nearby Utica Shale, to raise $900 million even though the bond offering was rated CCC+by Standard & Poor's (Loder 2014). Coupled with the fact that output from shale gas wells drops precipitously in the first year, shale gas producers maintain production levels by drilling more wells, which suggests more borrowing. The real question with funding seems to be whether this is a sustainable business model or not. It is up to the development of technology to drive the cost down and market demand to keep natural gas price stable. With the very low starting point of natural gas price in the USA which is about 20–25 % of the price in Europe and China, the potential profit margin is huge which attracts many large energy companies to rush to have a piece of the pie.[9] It also encourages more energy-intensive manufacturing companies to bring their operations back to the USA from overseas.

What Does Policy That Balances These Impacts Look Like?

Given that so much scientific research on shale gas development is being conducted by geologists, chemists, economists, and policy analysts, any policy recommendations at this time may be premature. States like New York have a moratorium on fracking at this time, presumably to await further research into the effects of fracking on the environment. However, we can start the dialogue about optimal policy mixes as they relate to fracking by looking at the results of a recent public opinion study by Brown et al. (2013). This study reports the results of surveys administered to residents of Pennsylvania and Michigan. Not surprising, "respondents view the economic benefits of hydraulic fracturing in their respective states as one of the primary benefits of fracking in the US" (p. 15). Also not surprising were the favorable perceptions of fracking along political party lines; Republicans generally had a more favorable view of fracking than Democrats. However, what was revealing is that "both Michigan and Pennsylvania respondents overwhelmingly agree

[9] "Williams Strikes Nearly $6 Billion Deal to Expand Shale Oil, Gas Holdings." http://online.wsj.com/articles/williams-strikes-nearly-6-billion-deal-to-expand-into-shale-oil-natural-gas-1402865062 (Retrieved on 07/24/2014)

that natural gas should be viewed as a public resource" (p. 18). This illustrates a fundamental question that must be answered before any policy prescriptions can be made – who owns the natural gas in question. This question of ownership hinges on whether property owners, who have rights to the land, also have rights to the air above and the subsurface below the land. This idea traces at least as far back as English common law in the case Bury v. Pope (1587), as described in Schick (1961). Harold Demsetz (1967) wrote:

> Property rights are an instrument of society and derive their significance from the fact that they help a man form those expectations which he can reasonably hold in his dealings with others. These expectations find expression in the laws, customs, and mores of a society. An owner of property rights possesses the consent of fellowmen to allow him to act in particular ways.

Cole and Grossman (2002) point out that a system of property rights forms the basis of all market exchange:

> .... the allocation of property rights in society affects the efficiency of resource use. More generally, assumptions of well-defined property rights underlie all theoretical and empirical research about functioning markets. The literature further assumes that when rights are not clearly defined, market failures result. The meaning of property rights is, thus, central to the language of economics.

Property rights (i.e., ownership) are a major issue that takes precedence over zoning, taxation, and chemical disclosures related to shale gas development because ownership of the natural gas determines who is legally able to give permission to extract the gas, occupy one's land to set up drilling wells, and sell the extracted gas on the open market. The conflict between private property rights and public interest is not a new question. In early 1970s, public interests in environmental quality grew with the establishment of EPA. People were puzzled by questions such as how to balance automobile pollution and clean air and whether individual right of driving should be banned to protect public interest of clean air. Eventually, there were more environmentally friendly cars produced with high energy efficiency, and people still drive today. Sax (1971) discussed these kinds of social dilemma and stated that "... rather than fumbling with doctrine labels and legal accusations, we can put our energy into trying to determine what resolution of conflicting uses is likely to maximize total net benefits for us, and how we can best achieve that goal." This is a wise suggestion and very relevant to today's debate on shale gas development.

Future Research

How municipalities and states plan for and adjust tax policies in order to maximize the benefits received from fracking activities is an important area for additional research. In like manner, how local communities plan to and have successfully attracted oil/gas workers from other areas is a highly relevant topic.

How the increase in demand for natural gas impacts local infrastructure (e.g., wastewater treatment and transportation, vehicle transportation) is another area where research would be useful. There is a lot of research about how many jobs are created and how many tax dollars are created, but not much on the costs municipalities incur. If we are looking at checks and balances as any good accountant would, we should know both the revenues and costs in order to determine if a project is viable from a cost/benefit perspective.

Research on the cost associated with any known environmental events that have occurred as a results of fracking is important to understanding what the real risks associated with these activities are. At present we can look to other types of contamination events as an indicator, but empirical numbers from today's market would be helpful in the planning process. If, for instance, municipalities knew on a well-by-well basis the range of damages they would be dealing with if a contamination event occurred, tax dollars could be allocated to a fund based on the number of wells permitted. This is just one example. In addition, insurance companies might be able or willing to insure against the risk (real or perceived) of contamination if they understood the dollar figures they would be dealing with should such an event occur.

Also, future research will not be limited to the USA. Recently lots of discussion has occurred on whether or not to eliminate the export ban on oil that the USA currently has in effect (Brown et al. 2014). Relatedly, shale gas development in the USA will impact world gas markets. Situations like the proposed liquefied natural gas [LNG] terminals being developed in coastal cities (for instance the future Jordan Cove LNG Terminal in Oregon, which was approved earlier this year by the US Department of Energy[10]) will have domestic and international impacts and are a result of the USA potentially becoming a net exporter of natural gas (i.e., the Energy Information Administration estimates this will occur before the year 2020).

References

Adams R, Kelsey TW (2012) Pennsylvania dairy farms and Marcellus shale, 2007–2010. Marcellus Education Fact Sheet. Penn State Cooperative Extension, University Park

Boxall PC, Chan WH, McMillan ML (2005) The impact of oil and natural gas facilities on rural residential property values: a spatial hedonic analysis. Resour Energy Econ 27:248–269

Brown E, Hartman K, Borick C, Rabe BG, Ivacko T (2013) The national surveys on energy and environment. Public opinion on fracking: perspectives from Michigan and Pennsylvania. University of Michigan, The Center for Local, State, and Urban Policy, Gerald R. Ford School of Public Policy (May)

Brown SPA et al (2014) Crude behavior: how lifting the export ban reduces gasoline prices in the United States, Issue brief 14-03-REV. Resources for the Future, Washington, DC

[10] "Energy Department OKs LNG terminal on Oregon coast." http://seattletimes.com/html/businesstechnology/2023217400_apxlngexportsoregon.html (Retrieved on 07/27/2014)

Bureau of Economic Analysis (2014) Comparison of personal saving in the National Income and Product Accounts (NIPAs) with personal saving in the Flow of Funds Accounts (FFAs). http://www.bea.gov/national/nipaweb/nipa-frb.asp. Accessed 5 May 2014

Bury v. Pope (1587) Cro. Eliz. 118 [78 Eng. Rep. 375]

Cole DH, Grossman PZ (2002) The meaning of property rights: law versus economics? Land Econ 78(3):317–330

Demsetz H (1967) Toward a theory of property rights, the American economic review, 57:2, papers and proceedings of the 79th annual meeting of the American Economic Association, pp 347–359.

Finkel ML, Selegean J, Hays J et al (2013) Marcellus shale drilling's impact on the dairy industry in Pennsylvania: a descriptive report. New Solut 23(1):189–201

Frey J (2012) The future of the Pennsylvania dairy industry with the impact of natural gas. Center for Dairy Excellence, Harrisburg

Gelber M (2014) What it takes to strike job gold in North Dakota's oil boom. http://jobs.aol.com/articles/2014/03/20/oil-boom-north-dakota-what-it-takes/. Accessed 21 May 2014

Gopalakrishnan S, Klaiber HA (2012) Is the shale boom a bust for nearby residents? Evidence from housing values in Pennsylvania. Working paper, The Ohio State University, Columbus

Hefley WE et al (2011) The economic impact of the value chain of a Marcellus shale well, Pitt business working papers. University of Pittsburgh Katz Graduate School of Business, Pittsburgh

Higginbotham A et al (2010) The economic impact of the natural gas industry and the Marcellus Shale development in West Virginia in 2009. Bureau of Business and Economic Research College of Business and Economics West Virginia University, Morgantown, West Virginia

Kelsey TW et al (2011) Economic impacts of Marcellus shale in Pennsylvania: employment and income in 2009. Marcellus Shale Education & Training Center (a collaboration of Pennsylvania College of Technology and Penn State Extension), State College, Pennsylvania

Kelsey TW et al (2012a) Economic impacts of Marcellus shale in Bradford county: employment and income 2010. Marcellus Shale Education & Training Center (a collaboration of Pennsylvania College of Technology and Penn State Extension), State College, Pennsylvania

Kelsey TW et al (2012b) Economic impacts of Marcellus shale in Sullivan county: employment and income 2010. Marcellus Shale Education & Training Center (a collaboration of Pennsylvania College of Technology and Penn State Extension), State College, Pennsylvania

Kelsey TW et al (2012c) Economic impacts of Marcellus shale in Susquehanna county: employment and income 2010. Marcellus Shale Education & Training Center (a collaboration of Pennsylvania College of Technology and Penn State Extension), State College, Pennsylvania

Kelsey TW et al (2012d) Economic impacts of Marcellus shale in Tioga county: employment and income 2010. Marcellus Shale Education & Training Center (a collaboration of Pennsylvania College of Technology and Penn State Extension), State College, Pennsylvania

Kelsey TW et al (2012e) Economic impacts of Marcellus shale in Wyoming county: employment and income 2010. Marcellus Shale Education & Training Center (a collaboration of Pennsylvania College of Technology and Penn State Extension), State College, Pennsylvania

Loder A (2014) Junk bonds fuel the shale boom. Bloomberg BusinessWeek, 5 May 2014, pp 45–46

Muehlenbachs L, Spiller E, Timmins C (2014) The housing market impacts of shale gas development. Resources for the future discussion paper 13-39-REV, Jan, Washington, DC

National Academy of Sciences (2012) Induced seismicity potential in energy technologies. The National Academies Press, Washington, DC

Nationwide Mutual Insurance Company (2012) Nationwide statement regarding concerns about hydraulic fracturing, press release dated 13 July 2012. http://www.nationwide.com/newsroom/071312-FrackingStatement.jsp. Accessed Jan 2013

O'Donnell JK (2014) Where the oil boom is making airports zoom. Bloomberg BusinessWeek, 28 Apr 2014, pp 27–28

Oyakawa J et al (2013) Economic impact of the Eagle Ford Shale. Center for Community and Business Research, Institute for Economic Development, The University of Texas as San Antonio, San Antonio, Texas

Radow EN (2011) Homeowners and gas drilling leases: boon or bust. New York State Bar Association Journal (Nov/Dec)

Sax JL (1971) Takings, private property and public rights. Yale Law J 81(2):149–186

Schick FB (1961) Space law and space politics. Int Comp Law Q 10(4):681–706

Wobbekind R, Lewandowski B (2014) Hydraulic fracturing ban the economic impact of a statewide fracking ban in Colorado. The Business Research Division, Leeds School of Business, The University of Colorado Boulder

Wood G (2013) Williston, ND, Oil boom means rents as high as New York City's. http://realestate.aol.com/blog/2013/01/14/williston-north-dakota-oil-boom-rental-costs/. Accessed 21 May 2014

CPSIA information can be obtained at www.ICGtesting.com
Printed in the USA
LVOW01*1030010215
425224LV00001B/36/P

9 783319 114989